OTOLARYNGOLOGY RESEARCH ADVANCES

SALIVARY GLANDS

STRUCTURE, FUNCTIONS AND REGULATION

OTOLARYNGOLOGY RESEARCH ADVANCES

Additional books and e-books in this series can be found on Nova's website under the Series tab.

OTOLARYNGOLOGY RESEARCH ADVANCES

SALIVARY GLANDS

STRUCTURE, FUNCTIONS AND REGULATION

ALICIA S. BRYANT
EDITOR

Copyright © 2020 by Nova Science Publishers, Inc.

All rights reserved. No part of this book may be reproduced, stored in a retrieval system or transmitted in any form or by any means: electronic, electrostatic, magnetic, tape, mechanical photocopying, recording or otherwise without the written permission of the Publisher.

We have partnered with Copyright Clearance Center to make it easy for you to obtain permissions to reuse content from this publication. Simply navigate to this publication's page on Nova's website and locate the "Get Permission" button below the title description. This button is linked directly to the title's permission page on copyright.com. Alternatively, you can visit copyright.com and search by title, ISBN, or ISSN.

For further questions about using the service on copyright.com, please contact:
Copyright Clearance Center
Phone: +1-(978) 750-8400   Fax: +1-(978) 750-4470   E-mail: info@copyright.com.

NOTICE TO THE READER

The Publisher has taken reasonable care in the preparation of this book, but makes no expressed or implied warranty of any kind and assumes no responsibility for any errors or omissions. No liability is assumed for incidental or consequential damages in connection with or arising out of information contained in this book. The Publisher shall not be liable for any special, consequential, or exemplary damages resulting, in whole or in part, from the readers' use of, or reliance upon, this material. Any parts of this book based on government reports are so indicated and copyright is claimed for those parts to the extent applicable to compilations of such works.

Independent verification should be sought for any data, advice or recommendations contained in this book. In addition, no responsibility is assumed by the Publisher for any injury and/or damage to persons or property arising from any methods, products, instructions, ideas or otherwise contained in this publication.

This publication is designed to provide accurate and authoritative information with regard to the subject matter covered herein. It is sold with the clear understanding that the Publisher is not engaged in rendering legal or any other professional services. If legal or any other expert assistance is required, the services of a competent person should be sought. FROM A DECLARATION OF PARTICIPANTS JOINTLY ADOPTED BY A COMMITTEE OF THE AMERICAN BAR ASSOCIATION AND A COMMITTEE OF PUBLISHERS.

Additional color graphics may be available in the e-book version of this book.

Library of Congress Cataloging-in-Publication Data

ISBN: 978-1-53617-497-7

Published by Nova Science Publishers, Inc. † New York

CONTENTS

Preface vii

Chapter 1 Role of MicroRNAs and Long Noncoding RNAs
in Salivary Gland Physiology and Pathology 1
*Emanuela Boštjančič, Nina Hauptman
and Metka Volavšek*

Chapter 2 Parotid Gland Neoplasm:
Diagnosis and Management 35
Salma M. Al Sheibani

Chapter 3 Juvenile Recurrent Parotitis 97
Salma M. Al Sheibani

Index 113

Related Nova Publications 119

PREFACE

Salivary Glands: Structure, Functions and Regulation summarizes the current understanding of the critical role that miRNAs and lncRNAs may play in salivary gland development and physiology, tumorigenesis and its progression, as well as in other pathologies. The authors also evaluate their potential as circulating biomarkers and therapeutic targets in different pathologies.

Next, the authors explore the diagnostic assessment, surgical approaches, and use of radiation and chemotherapy in the treatment of parotid tumors. Furthermore, the most common complications of parotid surgery and relative management are considered.

The clinical presentation, etiology and diagnostic modality of juvenile recurrent parotitis is reviewed, in conjunction with the role of sialoendoscopy in the diagnosis, management and prevention of recurrent attacks.

Chapter 1 - About 2% of transcribed eukaryotic genomes are protein coding, while the vast majority of transcripts belong to non-coding RNAs (ncRNAs). ncRNAs are functional transcripts that do not code for proteins and are predominantly classified as (i) long non-coding RNAs (lncRNAs), which are >200 nt long, and (ii) small double- or single-stranded RNAs (dsRNAs), of which microRNAs (miRNAs) are the most widely studied and characterized of all the ncRNAs. ncRNAs in general are important regulators of gene expression in many eukaryotes and show greater tissue specificity and unique spatial and temporal expression patterns compared to protein-coding mRNAs. This knowledge makes them attractive in the search of novel diagnostics/prognostics cancer biomarkers in body fluid samples. Genome-wide profiling has shown that ncRNAs have distinct signatures specific for a certain cancer type. Moreover, global mean expression profiling has in several studies demonstrated differential expression of panels of numerous ncRNAs suggesting that they may be useful biomarkers for accurate classification of cancer. Both, miRNAs and lncRNAs have been implicated as having tumor suppressor and oncogenic roles. In this chapter, the authors are summarizing the current understanding of the critical role that miRNAs and lncRNAs may play in salivary gland development and physiology, tumorigenesis and its progression as well as in other pathologies (e.g., Sjögren syndrome). The authors are also evaluating their potential as circulating biomarkers (serum, saliva) and therapeutic targets in different pathologies. The authors are summarizing their expression in experimental models (cell lines, animal studies, and human tissue) in terms of genome-wide and cell- and tissue-type specific function. The special emphasis is placed on salivary gland tumors, since prognosis for patients harbouring malignant tumors remains poor across all stages, despite advances in staging techniques and treatments. miRNAs and other ncRNAs expression profiles in salivary gland tumors and other diseases highlighted the potential value of this class of RNAs as markers in patient diagnosis and prognosis.

Chapter 2 - The parotid gland is the largest of the major salivary glands and the most common site for salivary tumors, the majority of which are benign. Parotid gland tumors encompass a mixture of different histologies that can be categorized into low, moderate, and high risk classifications based on their unique biologies and clinical outcomes. Fine needle aspiration should be considered as a part of the diagnostic assessment, but it has varying sensitivities and specificities. Anatomic imaging of the parotid glands, such as ultrasound, CT scans, and MRI, is crucial for planning of surgical management.

Surgery is the primary modality for management of these tumors. Facial nerve monitoring can be selectively used to decrease the morbidity associated with this type of surgery. The complications of parotid surgery can be classified into intra-operative and post-operative. The latter type can be subdivided into early and late complications. This operation continues to be a challenge on account of the wide range of tumors found and the differences in size and location. It is, therefore, mandatory that an experienced surgeon performs this operation.

Low risk salivary gland tumors are often treated with adequate surgical resection and do not require adjuvant radiation because they display very low rates of recurrence. Adjuvant radiation therapy is required in moderate risk salivary gland tumors as local recurrence can occur frequently even after optimal surgical management. High risk salivary gland tumors carry a substantial risk of spread to the ipsilateral neck nodes and show an improvement in local control and survival with adjuvant radiation therapy. Progress in modern radiation techniques now permit treatment of salivary gland tumors with fewer adverse effects and excellent local and regional control. Chemotherapy remains to have a palliative role in the management of salivary gland tumors, though researchers in this field are trying to identify a therapeutic role for chemotherapy in order to improve overall survival.

This chapter reviews parotid tumors, diagnostic assessment, surgical approaches, use of neck dissection, and use of radiation and chemotherapy. In addition, the management of the facial nerve in malignant parotid tumors is highlighted. Furthermore, the most common complications of parotid surgery and relative management are taken into consideration.

Chapter 3 - Juvenile recurrent parotitis is the second most common salivary gland disease in children after mumps. It's more common in males and in the age group between 3 and 6 years. It is characterized by recurrent attacks of non-suppurative and non-obstructive parotid inflammation. The pathogenesis is multifactorial and therefore the management is a challenge due to etiological diversity.

The main criterion for establishing the severity is the frequency of attacks. The treatment of the acute phase is with antibiotics and analgesics. Juvenile recurrent parotitis usually resolves spontaneously after puberty, however, in some cases the disease may continue leading to progressive loss of parenchymal functions. Subsequently, it may lead to a major intervention such as parotidectomy.

The diagnosis is established by history and clinical examination. Various imaging modalities are also used such as ultrasound, sialography and MR sialography.

The main difficulty of treatment of this disease concerns the prevention of the recurrence of inflammatory episodes.

Various medical and surgical measures have been used, but none of them proved to be useful in preventing or treating the attacks of juvenile recurrent parotitis. Sialoendoscopy is used as a diagnostic and therapeutic modality in juvenile recurrent parotitis. The main aim of this treatment is to reduce the recurrent attacks of parotitis and prevent irreversible changes in the parotid glands by irrigating and dilating the ductal system of the parotid gland under direct vision. Another advantage is the opportunity to inject medications intraductally under direct vision. Various intraductal lavage solutions were effective in prevention of recurrence regardless of their composition as it breaks the inflammatory cycle.

This chapter reviews the clinical presentation, etiology and diagnostic modality of juvenile recurrent parotitis. In addition to the review of acute juvenile recurrent parotitis attacks management, the role of sialoendoscopy in the diagnosis, management and prevention of recurrent attacks is addressed. Moreover, various solutions that can be used as a lavage solution and their role in the management of juvenile recurrent parotitis will be reviewed.

In: Salivary Glands
Editor: Alicia S. Bryant

ISBN: 978-1-53617-497-7
© 2020 Nova Science Publishers, Inc.

Chapter 1

ROLE OF MICRORNAS AND LONG NONCODING RNAS IN SALIVARY GLAND PHYSIOLOGY AND PATHOLOGY

Emanuela Boštjančič*, PhD, Nina Hauptman, PhD and Metka Volavšek, MD, PhD

Institute of Pathology, Faculty of Medicine,
University of Ljubljana, Ljubljana, Slovenia

ABSTRACT

About 2% of transcribed eukaryotic genomes are protein coding, while the vast majority of transcripts belong to non-coding RNAs (ncRNAs). ncRNAs are functional transcripts that do not code for proteins and are predominantly classified as (i) long non-coding RNAs (lncRNAs), which are >200 nt long, and (ii) small double- or single-stranded RNAs (dsRNAs), of which microRNAs (miRNAs) are the most widely studied and characterized of all the ncRNAs. ncRNAs in general are important regulators of gene expression in many eukaryotes and show greater tissue

---

* Corresponding Author Email: emanuela.bostjancic@mf.uni-lj.si.

specificity and unique spatial and temporal expression patterns compared to protein-coding mRNAs. This knowledge makes them attractive in the search of novel diagnostics/prognostics cancer biomarkers in body fluid samples. Genome-wide profiling has shown that ncRNAs have distinct signatures specific for a certain cancer type. Moreover, global mean expression profiling has in several studies demonstrated differential expression of panels of numerous ncRNAs suggesting that they may be useful biomarkers for accurate classification of cancer. Both, miRNAs and lncRNAs have been implicated as having tumor suppressor and oncogenic roles. In this chapter, we are summarizing the current understanding of the critical role that miRNAs and lncRNAs may play in salivary gland development and physiology, tumorigenesis and its progression as well as in other pathologies (e.g., Sjögren syndrome). We are also evaluating their potential as circulating biomarkers (serum, saliva) and therapeutic targets in different pathologies. We are summarizing their expression in experimental models (cell lines, animal studies, and human tissue) in terms of genome-wide and cell- and tissue-type specific function. The special emphasis is placed on salivary gland tumors, since prognosis for patients harbouring malignant tumors remains poor across all stages, despite advances in staging techniques and treatments. miRNAs and other ncRNAs expression profiles in salivary gland tumors and other diseases highlighted the potential value of this class of RNAs as markers in patient diagnosis and prognosis.

Keywords: miRNAs, lncRNAs, salivary glands, salivary gland tumors, Sjögren's syndrome.

1. INTRODUCTION

In the past, the central dogma of molecular biology centered on protein-coding genes and their proteins as molecules with cellular functions. Now, with improvement and introduction of new technologies, evidence is emerging that at least 90% of human genome is actively transcribed. The majority of transcripts are noncoding RNAs (ncRNAs). ncRNAs are a complex group of functional transcripts that do not code for proteins and are predominantly classified as (i) long non-coding RNAs (lncRNAs), which are >200 nucleotides long, and (ii) small double- or single-stranded RNAs (dsRNAs), of which microRNAs (miRNAs) are the most widely studied and characterized of all the ncRNAs (Esteller 2011).

miRNAs are much easier to investigate due to their higher stability, especially in complex tissue samples, like clinical samples. It is estimated that there could be as many as several thousands of miRNAs, thought to regulate 30-90% of genes within the human genome. miRNAs are important regulators of gene expression in many eukaryotes and show greater tissue specificity and unique spatial and temporal specific expression patterns compared to protein-coding mRNAs. This knowledge makes them attractive in the search of novel diagnostics/prognostics cancer biomarkers in body fluid samples. Genome-wide profiling has shown that miRNAs have distinct signatures specific for a certain cancer type. Moreover, global mean expression profiling has in several studies demonstrated differential expression of panels of numerous miRNAs suggesting that they may be useful biomarkers for accurate classification of cancer. miRNAs have been implicated as having tumor suppressor and oncogenic roles (Ying, Chang, and Lin 2008).

lncRNAs are transcripts lacking open-reading frame. Based on genome location, lncRNA can be divided into five categories: (1) sense, when overlapping one or more exons of another transcript on the same strand; (2) antisense, when overlapping one or more exons of another transcript on the opposite strand; (3) bidirectional, when the sequence is located on the opposite strand from a neighboring coding transcript, whose transcription is initiated less than 1000 base pairs away; (4) intronic, when it is transcribed from an intron of a second transcript, or (5) intergenic, when it lies between two genes (Wang and Chang 2011).

Several functions of lncRNA have been discovered, such as signal, decoy, scaffold, guide, enhancer RNAs, and short peptides. In cancer, several mechanisms including active lncRNA molecules have been researched, mostly as chromatin remodeling, chromatin interactions, and natural antisense transcripts (NATs). lncRNA can act as signal lncRNA, scaffold lncRNA or as an enhancer RNA by interacting with gene promoters through cromatin looping. lncRNA can regulate transcription through acting as a sponge for regulatory factors, including transcription factors and catalytic proteins or subunits of larger chromatin-modification complexes. Complexes of lncRNA and miRNA are known as ceRNAs. NATs are believed to regulate the expression of sense transcripts, with which they overlap (Wang and Chang 2011). Several lncRNA have been studied in various types of cancers, but only a few have been reported in salivary gland malignancies.

In this chapter, we are summarizing the current understanding of the critical role that miRNAs and lncRNAs may play in salivary gland development and physiology, tumorigenesis and its progression as well as in other pathologies (e.g., Sjögren syndrome). We are evaluating their potential as circulating biomarkers (serum, saliva) and therapeutic targets, and summarizing their expression in experimental models (cell lines, animal studies) and human tissue in terms of genome-wide and cell- and tissue-type specific function. However, the detailed description of miRNAs and lncRNAs biogenesis, extracellular vesicles, especially exosomes, as well as salivary gland tumors and Sjögren syndrome is too complex and as as such beyond the scope of current review and can be found elsewhere (Wang and Chang 2011, Esteller 2011, Ying, Chang, and Lin 2008, Simpson et al. 2014, Gallo et al. 2012, Zhan et al. 2019).

2. SALIVARY GLAND DEVELOPMENT AND PHYSIOLOGY

2.1. Extracellular Vesicle Transfer

Extracellular vesicles (EVs, including exosomes) are a group of heterogeneous nanometer-sized vesicles that are released by all types of cells and serve as functional mediators of cell-to-cell communication. This ability is primarily due to their capacity to package and transport various proteins, lipids and nucleic acids as are miRNAs. These contents can influence the function and fate of both recipient and donor cells. Significant amounts of miRNAs are detected in exosomes, but their function during fetal development is poorly understood. Epithelial-mesenchymal interactions are required to coordinate cell proliferation, patterning, and functional differentiation of multiple cell types in a developing organ. It was reported that mature mesenchymal-derived miRNAs in the fetal mouse salivary gland are loaded into exosomes, and transported to the epithelium where they influence progenitor cell proliferation. The exosomal miRNAs regulate epithelial expression of genes involved in DNA methylation in progenitor cells to influence morphogenesis. Thus, exosomal miRNAs are mobile genetic signals that cross tissue boundaries within an organ. These findings raise many questions about how miRNA signals are initiated to coordinate organogenesis and whether they are master regulators of epithelial-mesenchymal interactions. The development of therapeutic applications using exosomal miRNA for the regeneration of damaged adult organs is a promising area of research (Hayashi and Hoffman 2017, Hayashi et al. 2017).

2.2. Branching Morphogenesis and Submandibular Salivary Gland

Branching morphogenesis is an important developmental process for many organs, including the salivary glands. Whereas epithelial-mesenchymal interactions, defined as cell-to-cell communications, are known to drive branching morphogenesis, the molecular mechanisms responsible for those inductive interactions are still largely unknown. Cell growth factors and integrins are known to be regulators of branching morphogenesis of salivary glands. In addition, functional miRNAs have recently been reported to be present in the developing submandibular gland. In this review, the authors describe the roles of various cell growth factors, integrins and miRNAs in branching morphogenesis of developmental mouse submandibular glands (Kashimata and Hayashi 2018).

Fetal murine submandibular salivary gland (SMG) is known as a model to study organogenesis including branching morphogenesis. Branching morphogenesis is a crucial developmental process in which vertebrate organs generate extensive epithelial surface area while retaining a compact size. In the vertebrate SMG, branching morphogenesis is crucial for the generation of the large surface area necessary to produce sufficient saliva. Forty-four miRNAs and novel miRNA candidates were detected in SMG suggesting that these miRNAs were associated with regulating organogenesis, possibly including branching morphogenesis (Hayashi et al. 2009).

Branching morphogenesis in murine SMG is also regulated by extracellular matrix (ECM) and many other biological processes through interactions between the epithelium and the mesenchyme. miR-21 expression in the mesenchyme was up-regulated and accelerated by epidermal growth factor, which is known to enhance branching morphogenesis in vitro. Down-regulation of miR-21 in the mesenchyme was associated with a decrease in the number of epithelial buds. Relative quantification of candidates for target genes of miR-21 indicated that two messenger RNAs, Reck and Pdcd4, were down-regulated in the mesenchyme, where miR-21 expression levels were up-regulated. These results suggest that branching morphogenesis is regulated by miR-21 through gene expression related to ECM degradation in the mesenchyme (Hayashi et al. 2011).

Using microarrays, miRNA expression profiles have been established at selected times during development of the murine first molar mandibular tooth germ and the right SMG. In tooth germ and salivary gland up to 88 different miRNAs were detected. miRNA expression was highly dynamic; miRNA profiles were changing extensively with time of development. Additionally, the expression of some miRNAs was tissue-specific. Bioinformatics analysis of clusters of miRNAs suggested miRNAs to be involved in the regulation of essential developmental processes, as are epithelial cell proliferation, mesodermal cell fate determination and salivary gland morphogenesis (Jevnaker and Osmundsen 2008).

The regulation of epithelial proliferation during organ morphogenesis is crucial for normal development, as dysregulation is associated with tumor formation. miRNAs, such as miR-200c, are post-transcriptional regulators of genes involved in cancer. However, the role of miR-200c during normal development is unknown. In the mouse developing SMG, miR-200c accumulates in the epithelial end buds. Using both loss- and gain-of-function, it was demonstrated that miR-200c reduces epithelial proliferation during SMG morphogenesis trough targeting the very low density lipoprotein receptor (Vldlr) and its ligand reelin. miR-200c thus influences epithelial proliferation during branching morphogenesis via a Vldlr-dependent mechanism. miR-200c and Vldlr may be novel targets for controlling epithelial morphogenesis during glandular repair or regeneration (Rebustini et al. 2012).

2.3. Salivary Acinar Cells

The transcription factor networks that drive parotid salivary gland progenitor cells to terminally differentiate, remain largely unknown. Therefore, mRNA and miRNA expression was measured across the month long process of differentiation in the parotid gland of the rat. Acinar cells were isolated at either nine time points (mRNA) or four time points (miRNA) using laser capture microdissection. Global mRNA and miRNA expression was measured at different ages and 2656 mRNAs and 64 microRNAs were identified as differentially expressed. Because mRNA expression was sampled at many time points, clustering and regression analysis were able to identify dynamic expression patterns that had not been implicated in acinar differentiation before. Further clustering of microarray measurements suggests that expression occurs in four stages. Seventy-nine miRNAs are significantly differentially expressed across time. Reciprocal correlations of expression of miRNAs and their target mRNAs, suggest a putative network involving Klf4 and several targeting miRNAs. The network suggests a molecular switch (involving Prdm1, Sox11, Pax5, Xbp1, miR-200a, miR-30a and miR-214). Network analysis thus identified a novel stemness genetic switch involving transcription factors and miRNAs, and transition to an Xbp1 driven differentiation network (Metzler, Venkatesh, et al. 2015, Metzler, Appana, et al. 2015).

2.4. Autophagy

Autophagy is a lysosome-dependent degradation process that is involved in both cell survival and cell death, but little is known about the mechanisms that distinguish its use during these distinct cell fates. miR-14 has been identified as being both necessary and sufficient for autophagy during developmentally regulated cell death in Drosophila. Mis-expression of miR-14 was sufficient to prematurely induce autophagy in salivary glands. miR-14 was also shown to regulate inositol 1,4,5-trisphosphate kinase 2 (ip3k2), affecting inositol 1,4,5-trisphosphate (IP3) signaling and calcium levels during salivary gland cell death (Nelson, Ambros, and Baehrecke 2014).

3. SALIVARY GLAND TUMORIGENESIS AND ITS PROGRESSION

Salivary gland tumors (SGTs) are rare tumors of the head and neck with different clinical behavior. Histologically and clinically SGTs are heterogeneous group of tumours with missing prognostic factors and therapeutic targets. Preoperative diagnosis is crucial for their correct management. The identification of molecular markers might improve the accuracy of pre-surgical diagnosis. They account for about 3% of all head and neck tumors (Simpson et al. 2014).

Our previous study included 70 SGTs of different histological subtypes. We have analysed expression of miR-99b, miR-133b, miR-140, miR-140-3p, and let-7a in correlation to immunohistochemical expression and copy number variation (CNV) of EGFR. Based on histological subtypes, we found differential expression of all five miRNAs. We confirmed association of reactivity of EGFR, miR-133b, miR-140, miR-140-3p, and let-7a with CNV of EGFR and a positive association between miR-133b/let-7a and reactivity of EGFR (Bostjancic et al. 2017).

Another study investigated the expression of a panel of 798 miRNAs using Nanostring technology in 14 patients with malignant SGTs (mucoepidermoid carcinomas or MECs, adenoid cystic carcinomas or ACCs, acinic cell carcinoma, salivary duct carcinoma or SDC, cystadenocarcinoma and adenocarcinoma) and in 10 patients with benign SGTs (pleomorphic adenomas, PA). Forty six miRNAs were differentially expressed between malignant and benign SGTs. Interestingly, clustering analysis showed that this signature of 46 miRNAs is able to differentiate the two analyzed groups suggesting a correlation between histological diagnosis (benign or malignant) and miRNA expression profile. The molecular signature identified in this study might become an important preoperative diagnostic tool (Denaro et al. 2019).

Another study also aimed to find out differential expression of 95 miRNA profiles between 20 patients with either benign or malignant SGTs. Tissue, serum and saliva samples were collected. When tissue samples were studied miR-21, miR-31, miR-199a-5p, miR-146b, miR-345 were up-regulated in the malignant group compared to benign group (Cinpolat et al. 2017).

There is significant controversy in the literature regarding the relationship between *hypoxia* and SGTs. In the study on a total of 62 samples, no differences in HIF-1α expression were observed among the control group, benign and malignant SGTs. Also, the angiogenic markers, miR-210 and HIF-1α, do not appear to distinguish malignancy in salivary glands (Cardoso et al. 2019).

Further, the expression profile of *angiogenesis-related* miRNAs has been evaluated in SGTs. Expression of miR-9, miR-16, miR-17, miR-132, miR-195 and miR-221 was assessed in 11 ACCs, 9 MECs and 11 PAs. miR-9, miR-16, miR-17, miR-132, miR-195 and miR-221 were dysregulated in most cases compared to normal tissues. miR-9 showed a statistical significant negative correlation with micro vessel density. miR-17, miR-132, miR-195 and miR-221 were suggested to play an important role as tumor suppressor in SGTs (Santos et al. 2017).

Apoptosis-related miRNAs (miR-15a, miR-16, miR-17-5p, miR-20a, miR-21, miR-29, and miR-34) and their target mRNAs were also analysed in 25 PAs, 23 MECs, and 10 non-neoplastic salivary gland samples. Upregulation of miR-15a, miR-16, miR-17-5p, miR-21, miR-29, and miR-34a was observed in PAs, whereas miR-21 and miR-34 were upregulated in majority of MECs and downregulation of miR-20a was observed in majority of PAs and MECs. Study provided evidence of alterations in the expression of apoptosis-regulating miRNAs in SGTs (Flores et al. 2017).

3.1. Benign Tumors

Genome-wide miRNA expression profiling of salivary gland PAs revealed a distinct expression signature consisting largely of upregulated miRNAs compared with matched normal tissue. Five miRNA, genes upregulated in the tumours, were found in close proximity to fragile sites and/or cancer associated genomic regions. The expression of components of the miRNA processing machinery (Dicer, Drosha, DGCR8 and p68) suggest that the deregulation of miRNA expression may result from increased miRNA biogenesis. This is the first study to examine changes in the miRNAs in PA, the most common SGTs. These changes may be potential underlying mechanisms for the development of these benign tumours (Zhang et al. 2009).

The high mobility group protein HMGA2 plays an important role as a chromatin component of stem cells and as a protein causally related to the development of a variety of benign tumors (e.g., PA of the salivary glands). A highly conserved region within intron 3 of HMGA2, encoding a miRNA, was described and co-expression with HMGA2 suggests that as an intronic miRNA, this miRNA may cooperate with HMGA2 in its physiological and/or aberrant functions (von Ahsen et al. 2008).

Hyperactivation of the Wnt/β-catenin pathway promotes tumor initiation, tumor growth and metastasis in various tissues. Although there is evidence for the involvement of Wnt/β-catenin pathway activation in SGTs, the precise mechanisms are unknown. Downregulation of the Wnt inhibitory factor 1 (WIF1) is a widespread event in salivary gland carcinoma ex-pleomorphic adenoma (CaExPA). It was shown that WIF1 downregulation occurs in the CaExPA precursor lesion PA and indicates a higher risk of progression from benign to malignant tumor. Furthermore, WIF1 significantly increased the expression of pri-let-7a and pri-miR-200c, negative regulators of stemness and cancer progression, and functions as a positive regulator of miR-200c (Ramachandran et al. 2014).

Another publication reported, by using in situ hybridization, significant overexpression of miR-181b in PAs, while miR-21, a known oncogenic miRNA, was undetectable in PAs and normal tissue (Andreasen et al. 2015).

3.2. Malignant Tumors

3.2.1. Adenoid Cystic Carcinoma (ACC)

ACC is among the most common SGTs, and is notorious for its unpredictable clinical course with frequent local recurrences and metastatic spread. However, the molecular mechanisms for metastatic spread are poorly understood. Therefore it is not surprising that this neoplasm is the most studied of the SGTs in the context of miRNAs (Simpson et al. 2014).

3.2.1.1. Investigation of Primary Tumors

One of the first studies investigated expression of selected miRNAs in a set of 30 primary ACCs, matched normal samples and a pooled salivary gland standard. Of the highly dysregulated miRNA in ACC, overexpression of the miR-17 and miR-20a were significantly associated with poor outcome in the screening and validation sets (Mitani et al. 2013).

miR-155 expression was further determined in ACC specimens along with normal salivary glands. The effect of miR-155 on tumor growth was also examined in vitro and in vivo using mouse models. MiR-155 was overexpressed in ACC, proliferation of ACC cells was markedly inhibited by knocking down miR-155, whereas inhibition of miR-155 significantly suppressed the invasive capacity of ACC cells. In vivo growth of ACC cell-derived tumors was significantly slower by inhibition of miR-155 suggesting oncogenic role for this miRNA (Liu et al. 2013).

Another miRNA, miR-125a-5p was down-regulated and closely related to the metastasis and progression in human ACC specimens. In vitro, miR-125a-5p overexpression can suppress ACC cell migration and invasion; while blocking miR-125a-5p can relieve inhibition effect. It directly targets p38 and tissue samples of patients indicated the negative correlation between miR-125a-5p and p38; clinical analysis further confirmed that low level expression of miR-125a-5p is closely associated with poor prognosis of ACC. Furthermore, down-regulation of miR-125a-5p triggered downstream p38/JNK/ERK activation. Taken together, these results indicate that down-regulation of miR-125a-5p promotes ACC progression through p38 signal pathway (Liang et al. 2017).

Using next-generation sequencing, miRNA and mRNA expression profiles were obtained from six paired primary salivary ACCs and corresponding adjacent normal glands. As much as 2107 significant differentially expressed mRNAs were identified as well as 40 differentially expressed miRNAs. Integrated analysis of miRNA and mRNA expression profiles recognized a core microRNA-mRNA regulatory network and unmasked many novel genes, identifying many mRNAs and miRNAs worthy further exploration in salivary ACC (Han et al. 2018).

The prognostic value of mRNAs was assessed in two ACC cohorts: a training cohort ($n = 64$) and a validation cohort ($n = 120$). Association of miRNAs with reduced recurrence-free survival and overall survival in a training cohort was not confirmed in the validation cohort. However, two distinct subsets of ACCs separated in miRNA expression, but without significant difference in outcome. Collectively, ACC is a heterogeneous group of malignancies in its miRNA profile (Andreasen, Tan, Agander, Hansen, et al. 2018).

3.2.1.2. Metastases Development

The mechanism involved in salivary ACC metastasis is not yet fully understood. The microarray data revealed that the levels of 38 miRNAs significantly differed between the ACC cells with high and low metastatic potential. The expression of miR-4487 and miR-4430 was significantly upregulated and expression of miR-5191 and miR-3131 significantly downregulated in the ACC cells with high metastsic potential (Chen et al. 2014).

Similarly, using Nanostring nCounter analysis, the miRNA expression in two salivary ACC cell lines with different metastasis rates was analysed. Higher expression of miR-130a, miR-342, and miR-205 was observed in low metastatic and higher expression of miR-99a and miR-155 in high metastatic cell lines. In human tissue, miR-205 was highly expressed in the primary ACC, while miR-155 and miR-342 were highly expressed in recurrent ACC. miR-99a, miR-155, miR-130a, miR-342, and miR-205 thus might play a role in metastasis of ACC (Feng et al. 2017).

Using high-throughput sequencing and ACC cell lines with different metastatic potential, it has been suggested that differentially expressed mRNAs and miRNAs are potentially involved in ACC metastasis and that miR-338-5p/3p target LAMC2 to impair motility and invasion of ACC (Wang, Zhang, Shi, et al. 2018).

Genetic and miRNA expressional landscape of primary ACC and corresponding metastatic lesions from 11 patients has been recently characterized. Global miRNA profiling identified several differentially expressed miRNAs between primary ACC and metastases as compared to normal salivary gland tissue. Interestingly, individual tumor pairs differed in miRNA profile, but there was no general difference between primary ACCs and metastases (Andreasen, Agander, et al. 2018).

One of the first examined miRNA in ACC cells was miR-181a, indicating that miR-181a influences ACC cell migration, invasion and proliferation in vitro, and it suppresses tumor growth and lung metastasis in vivo. Direct targeting of miR-181a to MAP2K1, MAPK1 and Snai2 was confirmed. These results indicate that miR-181a plays an important role in the metastasis of ACC (He et al. 2013).

Down-regulated miR-101-3p expression was detected in ACC tissues and ACC cell lines with high potential for metastasis. Ectopic expression of miR-101-3p significantly repressed the invasion, proliferation, colony formation, and formation of nude mice xenografts and induced potent apoptosis in ACC cell lines. Pim-1 oncogene was subsequently confirmed as a direct target gene of miR-101-3p in ACC. Functional restoration assays revealed that miR-101-3p inhibits cell growth and invasion by directly decreasing Pim-1 expression. Protein levels of Survivin, Cyclin D1 and β-catenin were also down-regulated by miR-101-3p. miR-101-3p enhanced the sensitivity of cisplatin in ACC cell lines. The novel miR-101-3p/Pim-1 axis may provide insights into the carcinogenesis and tumor progression of ACC (Liu et al. 2015).

miR-320a was down-regulated in high lung metastatic ACC cells compared with the corresponding low metastatic ACC, and inhibited adhesion, invasion and migration of ACC cells by targeting integrin beta 3 (ITGB3). In vivo, enforced miR-320a expression suppressed metastasis of ACC xenografts. In the two independent sets of samples from 302 and 148 patients, miR-320a was downregulated in primary ACCs with metastasis compared to those without metastasis, and low expression of this miRNA predicts poor patient survival and rapid metastasis. Multivariate analysis showed that miR-320a expression was an independent indicator of lung metastasis (Sun et al. 2015).

Regulation of miR-582-5p expression significantly influenced the migration, invasion and proliferation ability of salivary ACC cells by targeting FOXC1. In vivo, miR-582-5p overexpression suppressed the tumorigenesis and pulmonary metastasis of ACC. Lower expression of miR-582-5p expression in human tissue samples predicts unfavorable prognoses and high rates of metastasis (Wang et al. 2017).

miR-21 is highly expressed in a variety of tumors and has a role in promoting tumor development. The level of miR-21 expression was significantly higher in ACC than that in normal salivary tissues, and it is also higher in tumors with metastasis than those without metastasis. Using an anti-miR-21 inhibitor in an in vitro model, downregulation of miR-21 significantly decreased the capacity of invasion and migration of ACC cells, whereas a pre-miR-21 increased the capacity of invasion and migration of ACC cells. PDCD4, which has been implicated in invasion and metastasis was identified as a direct miR-21 target gene. The suppression of miR-21 in metastatic ACC cells significantly increased the report activity of PDCD4 promoter and the expression of PDCD4 protein. The level of miR-21 expression positively related to the expression of PDCD4 protein in ACC specimens. Deregulation of miR-21 has an important role in tumor growth and invasion by targeting PDCD4 (Jiang et al. 2015). ACC cells with high metastatic potential exhibited a significantly higher expression of miR-21 compared with ACC cells with a lower metastatic potential indicating that miR-21 may be a novel target for ACC therapy and provide a novel basis for the clinical treatment of ACC (Yan et al. 2018).

3.2.1.3. Comparison to other Tumors

The expression levels of 18 miRNAs was analysed in breast and salivary ACC and control of normal breast and salivary gland tissues. Increased expression of miR-17 and miR-20a was found in breast ACC compared with normal tissue, while the expression level of let-7b and miR-193b was lower in salivary ACC compared with normal. Expression of miR-23b and miR-27b differed between normal breast and normal salivary gland tissue, but not between ACCs. The potential target mRNAs CCND1 and BCL2 were identified as reported targets of let-7b, miR-193b, miR-17, and miR-20a. Expression of their corresponding proteins cyclin D1 and Bcl-2 was overexpressed in breast and salivary ACCs in comparison with corresponding normal tissues. Although no differences in miRNA levels were found between breast and salivary ACCs, in both organs, miRNA expression level was significantly different between tumor and control tissue. Similar approach resulted in identifying 7 miRNAs overexpressed in salivary ACC cases and down-regulated in breast ACC (let-7b, let-7c, miR-17, miR-20a, miR-24, miR-195, miR-768-3), while 9 microRNAs were down-regulated in salivary ACC cases and overexpressed in breast ACC (let-7e, miR-23b, miR-27b, miR-193b, miR-320a, miR-320c, miR-768-5p, miR-1280 and miR-1826) relative to their controls. There were eight miRNAs, which were only expressed in salivary ACCs and one miRNA (miR-1234) which was only absent in salivary ACCs (Kiss, Tokes, Spisak, et al. 2015, Kiss, Tokes, Vranic, et al. 2015).

Additionally, functional annotation of the miRNAs differentially expressed between salivary gland and breast ACC showed that miRNA dysregulation is the first class of molecules separating ACC according to the site of origin (Andreasen, Tan, Agander, Steiner, et al. 2018).

Salivary ACC has a tendency to metastasize in lung or liver, without lymph node involvement, whereas squamous cell carcinoma of head and neck (HNSCC) preferentially metastasizes to regional lymph nodes. miRNA expression patterns of ACC and HNSCC were analysed in 21 tissue samples, resulting in five miRNAs, miR-214, miR-125a-5p, miR-574-3p, miR-199a-3p/199b-3p and miR-199a-5p to be over-expressed in ACC compared to HNSCC, whereas miR-452 showed a lower expression level. Different expression patterns of miRNA in ACC and HNSCC support the theory of tumor-specific expression and giving hints for different clinical behavior (Veit et al. 2015).

ACCs and polymorphous adenocarcinoma (PAC) are included among the most common SGTs. miR-150, miR-455-3p and miR-375 expression was analysed to identify a possible molecular distinction between ACC and PAC, which morphologically share many similarities. miR-150 and miR-375 was significantly decreased in both compared with salivary gland tissue controls, whilst miR-455-3p showed significantly increased expression in ACC when compared to PAC. Authors suggested that this miRNA could be used as a complimentary tool in the diagnosis of challenging ACC cases (Brown et al. 2019).

3.2.2. Mucoepidermoid Carcinoma (MEC)

MEC of salivary gland is a disease characterized by high rate of distant metastasis and associated with poor outcomes in high-grade cases. MEC is associated with translocation t(11;19), which results in CRTC1-MAML2 fusion and is considered an early event in oncogenesis of MEC. However, the molecular mechanisms underlying the MEC remain poorly understood (Simpson et al. 2014).

The TaqMan Human miRNA Cards Array was used for the miRNA profiling of MEC and normal tissues. miR-302a was the most significantly increased miRNA in cancer tissues and upregulation of miR-302a expression in SGT cell lines induced cancer cell invasion in vitro (Binmadi et al. 2018).

Identified through gene expression profiling analysis, lncRNA LINC00473, was the most down-regulated target in CRTC1-MAML2-

depleted human MEC cells. The LINC00473 transcription was significantly upregulated in CRTC1-MAML2-positive human MEC cell lines and primary MEC tumors and was highly correlated with the expression of CRTC1-MAML2 RNA. The CRTC1-MAML2 fusion activated CREB-mediated transcription, inducing LINC00473 transcription. When LINC00473 was depleted, reduction of proliferation and survival of human MEC cells in vitro was observed. Depletion of LINC00473 also prevented tumor growth in vivo in human MEC xenograft model. Cancer cell growth and survival genes are differentially expressed when LINC00473 is depleted, indicating the LINC00473 might regulate gene expression through binding to a component in cAMP signalling pathway (Chen et al. 2018).

Another study performed on MEC used the bioinformatics approach and identified differentially expressed ncRNAs by microarrays. A total of 3612 mRNA, 3091 lncRNAs, and 284 circular RNAs (circRNAs) were altered between MEC and normal tissue. The authors examined the co-expression networks between lncRNA and mRNA, and between circRNA and miRNA. Downregulation of circRNA hsa_circ_0012342 and upregulation of lncRNA NONHSAT154433.1 was validated with qPCR. Both ncRNA were related to pathogenesis of MEC (Lu et al. 2019).

3.2.3. Salivary Duct Carcinomas (SDCs)

SDC and Her2/Neu3-overexpressing invasive breast carcinomas (HNPIBC/IBC) are histologically indistinguishable. Common histopathologic and immune-phenotypic features of SDC and IBC are mirrored by a similar miRNA profile. MiRNA profiling of 5 SDCs, 6 IBCsHer2/Neu3+, and 5 high-grade ductal breast carcinoma in situ (DCIS) was performed by NanoString platform. Similar miRNA expression profiles were observed between IBC and SDC with the exception of 2 miRNAs, miR-10a and miR-142-3p, which were higher in IBCs. DCISs displayed increased expression of miR-10a, miR-99a, miR-331-3p and miR-335, and decreased expression of miR-15a, miR-16 and miR-19b compared to SDC. The normal salivary gland and breast tissues also showed similar expression profiles. Interestingly, miR-10a was selectively increased in both IBC and normal breast tissue compared to SDC and normal salivary gland tissue.

Downregulation of HOXA1, a miR-10 target, in IBC tumors compared to normal breast tissue was also observed. Taken together, data demonstrates that, based on miRNA profiling, SDC is closely related to HNPIBC. miR-10a is differentially expressed in IBC compared to SDC and may have potential utility as a diagnostic biomarker (Balatti et al. 2019).

4. SJÖGREN SYNDROME

Sjögren's syndrome (SS) is a complex autoimmune disease, which primarily affects salivary and lacrimal glands, causing loss of secretion resulting in the typical features of dry eyes (xerophtalmia) and mouth (xerostomia). Combining exocrine gland dysfunctions and lymphocytic infiltration, SS is characterised clinically by features of systemic autoimmunity and inflammation, with dysfunction of the exocrine glands, and histologically by lymphocytic infiltration, destruction and atrophy. While the pathogenesis of SS remains unclear, its etiology is multifunctional and includes a combination of genetic predispositions, environmental and epigenetic factors. Recently, interest has grown in the involvement of epigenetics in autoimmune diseases. Epigenetics is defined as changes in gene expression, that are inheritable and that do not entail changes in the DNA sequence. In SS, several epigenetic mechanisms are defective including DNA demethylation that predominates in epithelial cells, abnormal expression of miRNAs and lncRNAs, and abnormal chromatin positioning associated with autoantibody production. Last but not least, epigenetic modifications are reversible, as observed in MSGs from SS patients after B cell depletion using rituximab (Konsta et al. 2014).

4.1. Genome-Wide Studies

Comparative array analysis of miRNA expression in the salivary glands of SS and control subjects had revealed distinctive miRNA signatures in SS patients, associated with glandular inflammation and dysfunction. Furthermore, the expression analysis of certain miRNAs revealed their differential expression in the salivary gland tissues and peripheral blood mononuclear cells (PBMC) of SS patients. Although these association data implicate miRNAs in SS pathogenesis, thorough functional studies are needed to delineate their role in disease (Kapsogeorgou et al. 2011).

The ultra-deep sequencing of small RNAs was performed in patients with SS and healthy volunteers, primarily to identify and discover novel miRNA sequences that may play a role in the disease. The presence of six previously unidentified miRNA sequences in patient samples and in several cell lines was validated. One of the validated novel miRNAs showed promise as a biomarker for salivary function (Tandon et al. 2012).

Among 754 miRNA analyzed, 126 miRNA were identified that were significantly deregulated in SS compared to controls, with a trend that was inversely proportional with the impairment of salivary flow rates. Expression of different predicted miRNA-target genes was altered in SS patients with low salivary flow (Gallo et al. 2019).

4.2. Individual miRNAs

Differential expression of miR-17-92 cluster was investigated among varying histological stages of labial MSG in patients with primary SS. miR-17-92 cluster encodes 6 miRNAs, including miR-17, miR-18a, miR-19a, miR-19b, miR-20a and miR-92a. In the labial MSG, it was observed that the expression level of miR-18a was significantly up-regulated in patients with SS compared to healthy individuals, while the expression level of miR-92a was significantly downregulated. Also, no notable difference was observed in the expression levels of miR-17, miR-19a, miR-19b, and miR-20a. Furthermore, miR-18a was progressively up-regulated along the advanced histological states of the 3 SS subgroups, while the miR-92a was progressively downregulated. The association of increased expression levels of miR-18a and reduced expression levels of miR-92a with advanced clinical stages of SS could therefore significantly reduce the substantial subjectivity of scoring inflammatory infiltrates and may aid in the diagnosis of SS (Yan et al. 2019). Compared to healthy controls, downregulation of miR-1207-5p and miR-4695-3p expression was observed in the MSGs of primary SS patients (Yang et al. 2016).

4.3. miRNAs and lncRNAs

In patients with primary SS expression profiling by microarray was performed in labial salivary glands (LSGs). Totally 1243 lncRNAs and 1457 mRNAs were found deregulated. Eight of these lncRNAs were validated by qPCR and found significantly up-regulated in SS. Strong correlations were observed between these eight lncRNA and other clinical factors (Shi et al. 2016).

The interaction of lncRNAs-miRs exerts crucial functions in mediating inflammatory reaction. It is still unclear whether Mirt2-miR-377 mediates the inflammatory pathogenesis in SS. The inflammatory lesion model was established by stimulating salivary gland epithelial cells. The up-regulation of Mirt2 was observed in and Mirt2 overexpression restored the expression of miR-377. However, miR-377 silence abolished the protective effect on cell viability, inhibitory effect on apoptosis and prohibitive role in pro-inflammatory factors. Mirt2 diminished the phosphorylated expression of crucial regulators while miR-377 silence restored their phosphorylation. Mirt2 was elevated in investigated cells, leading to up-regulation of miR-377 in response to inflammatory lesions (Xin et al. 2019).

4.4. miRNAs Targets

The elevated tissue expression of Ro/SSA and La/SSB autoantigens appears to be crucial for the generation and perpetuation of autoimmune humoral responses against these autoantigens in SS. Gourzi et al. identified that let-7b, miR-16, miR-181a, miR-200b-3p, miR-200b-5p, miR-223 and miR-483-5p are predicted to target Ro/SSA and La/SSB mRNAs. To study possible associations with autoantigen, their expression was investigated in minor salivary gland (MSG) tissues, PBMC and long-term cultured non-neoplastic salivary gland epithelial cells from 29 SS patients and 24 controls. The levels of miR-16 were up-regulated in MSGs, miR-200b-3p in cells and miR-223 and miR-483-5p in PBMCs of SS patients compared to controls. The MSG levels of let-7b, miR-16, miR-181a, miR-223 and miR-483-5p were correlated positively with Ro-mRNAs, whereas let-7b, miR-200b-5p and miR-223 associated with La/SSB-mRNA. In PBMCs, let-7b, miR-16, miR-181a and miR-483-5p were correlated with Ro-mRNAs, whereas let-7b, miR-16 and miR-181a were also associated with La/SSB-mRNA expression. These findings indicate that miR-16, miR-200b-3p, miR-223 and miR-483-5p are deregulated in SS, but the exact role of this deregulation in disease pathogenesis and autoantigen expression needs to be elucidated (Gourzi et al. 2015).

miR-146a expression was analysed in PBMCs of 25 SS patients and ten healthy donors patients as well as in the SS-prone mouse model, to elucidate its involvement in SS pathogenesis. Expression of miR-146a was significantly increased in SS patients compared with healthy controls, and was upregulated in the salivary glands and PBMCs of the SS-prone mouse in both, prior to disease onset and in full-blown disease. More importantly, functional analysis revealed roles for miR-146a in increasing phagocytic activity and suppressing inflammatory cytokine production while migration, nitric oxide production and expression of antigen-presenting/costimulatory molecules are not affected in human monocytic cells. Taken together, data suggest that abnormal expression/regulation of miRNAs in innate immunity may contribute to or be indicative of the initiation and progression of SS (Pauley et al. 2011). In further research the same authors identified a target molecule of miR-146a and subpopulations of cells affected by altered miR-146a in the salivary glands of SS-prone mice. After in silico analyses, luciferase assay of the human CD80 3'-untranslated region demonstrated miR-146a directly inhibited CD80 protein expression. The specific reduction in CD80 protein was detected from the salivary gland epithelial cell population and in interstitial dendritic cells in the glands as well. The reduction in CD80 protein levels in salivary gland epithelial cells were negatively associated with elevated miR-146a expression (Gauna et al. 2015).

Salivary cystatin S is a defense protein mainly produced by submandibular glands and involved in innate oral immunity. miR-126 and miR-335-5p expression was analysed in MSG biopsies to verify whether an aberrant regulation of cystatin S at the glandular level may influence its salivary expression. Forty SS patients and 20 sex- and age-matched healthy volunteers were included. Salivary cystatin S was significantly decreased in SS patients compared to healthy volunteers, especially in those with hyposalivation, and the expression levels of miR-126 and miR-335-5p increased in inverse proportion. The mRNA of cystatin S did not change significantly, suggesting post-transcriptional regulation, suggesting that upregulation of miR-126 and miR-335-5p might be implicated in its expression (Martini et al. 2017).

miR-181a and miR-16 profile in labial salivary glands was analysed in SS patients and examined the correlation of levels with the pathological grade in SS. Subsequent to integrating all data base results, miR-181a and miR-16 were identified to be associated with the Ro/SS-associated antigen A and La/SS-associated antigen B during SS pathogenesis. miR-181a and miR-16 expression levels in the labial salivary gland of SS patients were decreased in comparison with those in the controls. Furthermore, the decreased expression levels of these miRNAs were associated with the labial salivary pathological focus scores, suggesting that miR-181a and miR-16 may serve a role in the pathogenesis of SS (Wang, Zhang, Zhang, et al. 2018).

5. CIRCULATING BIOMARKERS (SERUM, SALIVA)

Saliva is an easily accessible medium that has been shown to contain microvesicles (exosomes) that enclose miRNAs. It has been previously demonstrated that the majority of salivary miRNAs are within exosomes. miRNAs have been implicated in oral cancer and the use of salivary exosomal miRNAs holds the promise of identification of diagnostic and prognostic markers (Gallo and Alevizos 2013). Exosomes isolated from fresh and frozen glandular and whole human saliva were used as a source of miRNAs. The presence of miRNAs was validated with TaqMan quantitative PCR and miRNA microarrays. Healthy controls and a patient with SS were included. miRNAs extracted from the exosomal fraction were sufficient for qPCR and microarray profiling. The isolation of miRNAs from easily and non-invasively obtained salivary exosomes with subsequent characterization of the miRNA expression patterns is thus promising method for the development of future biomarkers of the diagnosis and prognosis of various salivary gland pathologies (Michael et al. 2010).

miRNAs 146a/b, miR-16, the miR-17-92 cluster and miR-181a were analysed in salivary and plasma samples, taken from primary SS patients with clinical, laboratory and ultrasound findings. SS patients had higher expression of salivary miR-146a than gender- and age-matched controls. The data showed that salivary miR-146a may represent a marker of the disease, and that the expression of salivary miR-17, miR-18a and miR-146b may be altered in patients with SS, and associated with worse clinical findings. Neither salivary nor plasma miRNAs correlated with disease duration or concomitant therapies (Talotta et al. 2019).

Whole saliva samples from patients with malignant (n = 38) or benign (n = 29) SGTs were obtained and used for miRNA profiling. Validation of selected miRNAs in an independent sample set was performed. With miRNA profiling, 57 of 750 investigated miRNAs were differently expressed, of which 54 showed higher miRNA expression in samples from patients with malignant tumors than those from patients with benign tumors. Validating the expression in an independent sample set of 9 miRNAs revealed higher expression of miRNAs in malignant samples compared with benign ones. The expression of 6 validated miRNAs was statistically significantly different between the two groups ($P < 0.05$). A four miRNA combination was able to discriminate between saliva samples from patients with malignant tumors from those of patients with benign parotid gland tumors (sensitivity 69%, specificity 95%) (Matse et al. 2013).

Another study also aimed to find out differential expression of 95 miRNA profiles between 20 patients with either benign or malignant SGTs. Serum and saliva samples were collected and miRNA expression was compared to those in tissue. Among studied miRNAs, miR-21, miR-23a, miR-27a, miR-223, miR-125b, miR-126, miR-146a, miR-30e were down regulated in the benign group compared to control group in the serum samples. miR-30e showed statistically significant up-regulation in malignant tumor group's plasma samples compared to benign group. However, there was no statistically significant difference in saliva samples between groups. Although there was no difference in saliva samples between groups, authors concluded that according to tissue and serum samples, miR-21 and miR-30e may have an important role; since they were down-regulated in benign tumors whereas up-regulated in malignant ones (Cinpolat et al. 2017).

6. MIRNAS AS THERAPEUTIC TARGETS

Extracellular vesicles (EVs, including exosomes) are a group of heterogeneous nanometer-sized vesicles that are released by all types of cells and serve as functional mediators of cell-to-cell communication. This ability is primarily due to their capacity to package and transport various proteins, lipids and nucleic acids as are miRNAs. These contents can influence the function and fate of both recipient and donor cells. More and more studies have shown that EVs are involved in every phase of cancer development, mediating bidirectional crosstalk between cancer cells and their tissue microenvironment. More specifically, EVs can promote tumor progression by modifying vesicular contents and establishing a distant pre metastatic niche with molecules that favor cancer cell proliferation, migration, invasion, metastasis, angiogenesis, and even drug resistance. Given that the packaging of these molecules is known to be tissue specific, EVs can serve as potential therapeutic targets and vehicles for drug delivery (Zhan et al. 2019).

Numerous studies showed dysregulation of miRNAs in ACC and SS, leading to suggestion that miRNAs might be a promising therapeutic targets. One example is oncogenic miR-21. Accordingly, while simvastatin (SIM) is effective against the growth of many cancer types and its side effects limit its use, effect of SIM in combination with miR-21 inhibitor against lung metastatic salivary ACC cells was analysed. It has been observed that such combination was effective in inhibiting progression of this cells, suggesting that multi-target therapy might represent a potentially successful approach in clinical treatment (Wang, Li, et al. 2018).

CONCLUSION

A lot of efforts have been made in the field of investigating miRNAs as prognostic, diagnostic or therapeutic targets in salivary gland pathologies, but there is still a lot to discover. Promising results of miRNAs research as target molecules and delivery vesicles raise from investigation of exosomes. However, regulation of biological functions and patahologic processes in salivary gland by the rest of ncRNAs (except miRNAs) is more or less unexplored and should be investigated in the future.

REFERENCES

Andreasen, S., T. K. Agander, K. Bjorndal, D. Erentaite, S. Heegaard, S. R. Larsen, L. C. Melchior, Q. Tan, B. P. Ulhoi, I. Wessel, and P. Homoe. 2018. "Genetic rearrangements, hotspot mutations, and microRNA expression in the progression of metastatic adenoid cystic carcinoma of the salivary gland." *Oncotarget* 9 (28):19675-19687. doi: 10.18632/oncotarget.24800.

Andreasen, S., Q. Tan, T. K. Agander, T. V. O. Hansen, P. Steiner, K. Bjorndal, E. Hogdall, S. R. Larsen, D. Erentaite, C. H. Olsen, B. P. Ulhoi, S. Heegaard, I. Wessel, and P. Homoe. 2018. "MicroRNA dysregulation in adenoid cystic carcinoma of the salivary gland in relation to prognosis and gene fusion status: a cohort study." *Virchows Arch* 473 (3):329-340. doi: 10.1007/s00428-018-2423-0.

Andreasen, S., Q. Tan, T. K. Agander, P. Steiner, K. Bjorndal, E. Hogdall, S. R. Larsen, D. Erentaite, C. H. Olsen, B. P. Ulhoi, S. L. von Holstein, I. Wessel, S. Heegaard, and P. Homoe. 2018. "Adenoid cystic carcinomas of the salivary gland, lacrimal gland, and breast are morphologically and genetically similar but have distinct microRNA expression profiles." *Mod Pathol* 31 (8):1211-1225. doi: 10.1038/s41379-018-0005-y.

Andreasen, S., M. H. Therkildsen, M. Grauslund, L. Friis-Hansen, I. Wessel, and P. Homoe. 2015. "Activation of the interleukin-6/Janus kinase/STAT3 pathway in pleomorphic adenoma of the parotid gland." *Apmis* 123 (8):706-15. doi: 10.1111/apm.12407.

Balatti, V., S. Oghumu, A. Bottoni, K. Maharry, L. Cascione, P. Fadda, A. Parwani, C. Croce, and O. H. Iwenofu. 2019. "MicroRNA profiling of salivary duct carcinoma versus Her2/Neu overexpressing breast carcinoma identify miR-10a as a putative breast related oncogene." *Head Neck Pathol* 13 (3):344-354. doi: 10.1007/s12105-018-0971-x.

Binmadi, N. O., J. R. Basile, P. Perez, A. Gallo, M. Tandon, W. Elias, S. I. Jang, and I. Alevizos. 2018. "miRNA expression profile of mucoepidermoid carcinoma." *Oral Dis* 24 (4):537-543. doi: 10.1111/odi.12800.

Bostjancic, E., N. Hauptman, A. Groselj, D. Glavac, and M. Volavsek. 2017. "Expression, mutation, and amplification status of EGFR and its correlation with five miRNAs in salivary gland tumours." *Biomed Res Int* 2017:9150402. doi: 10.1155/2017/9150402.

Brown, A. L., A. Al-Samadi, M. Sperandio, A. B. Soares, L. N. Teixeira, E. F. Martinez, A. P. D. Demasi, V. C. Araujo, I. Leivo, T. Salo, and F. Passador-Santos. 2019. "MiR-455-3p, miR-150 and miR-375 are aberrantly expressed in salivary gland adenoid cystic carcinoma and polymorphous adenocarcinoma." *J Oral Pathol Med*. doi: 10.1111/jop.12894.

Cardoso, C. M., S. F. de Jesus, M. G. de Souza, E. M. Santos, C. K. C. Santos, C. M. Silveira, S. H. S. Santos, A. M. B. de Paula, L. C. Farias, and A. L. S. Guimaraes. 2019. "Is HIF1-a deregulated in malignant salivary neoplasms?" *Gene* 701:41-45. doi: 10.1016/j.gene.2019.03.017.

Chen, W., X. Zhao, Z. Dong, G. Cao, and S. Zhang. 2014. "Identification of microRNA profiles in salivary adenoid cystic carcinoma cells during metastatic progression." *Oncol Lett* 7 (6):2029-2034. doi: 10.3892/ol.2014.1975.

Chen, Z., S. Lin, J. L. Li, W. Ni, R. Guo, J. Lu, F. J. Kaye, and L. Wu. 2018. "CRTC1-MAML2 fusion-induced lncRNA LINC00473 expression maintains the growth and survival of human mucoepidermoid carcinoma cells." *Oncogene* 37 (14):1885-1895. doi: 10.1038/s41388-017-0104-0.

Cinpolat, O., Z. N. Unal, O. Ismi, A. Gorur, and M. Unal. 2017. "Comparison of microRNA profiles between benign and malignant salivary gland tumors in tissue, blood and saliva samples: a prospective, case-control study." *Braz J Otorhinolaryngol* 83 (3):276-284. doi: 10.1016/j.bjorl.2016.03.013.

Denaro, M., E. Navari, C. Ugolini, V. Seccia, V. Donati, A. P. Casani, and F. Basolo. 2019. "A microRNA signature for the differential diagnosis of salivary gland tumors." *PLoS One* 14 (1):e0210968. doi: 10.1371/journal.pone.0210968.

Esteller, M. 2011. "Non-coding RNAs in human disease." *Nat Rev Genet* 12 (12):861-74. doi: 10.1038/nrg3074.

Feng, X., K. Matsuo, T. Zhang, Y. Hu, A. C. Mays, J. D. Browne, X. Zhou, and C. A. Sullivan. 2017. "MicroRNA Profiling and Target Genes Related to Metastasis of Salivary Adenoid Cystic Carcinoma." *Anticancer Res* 37 (7):3473-3481. doi: 10.21873/anticanres.11715.

Flores, B. C., S. V. Lourenco, A. S. Damascena, L. P. Kowaslki, F. A. Soares, and C. M. Coutinho-Camillo. 2017. "Altered expression of apoptosis-regulating miRNAs in salivary gland tumors suggests their involvement in salivary gland tumorigenesis." *Virchows Arch* 470 (3):291-299. doi: 10.1007/s00428-016-2049-z.

Gallo, A., and I. Alevizos. 2013. "Isolation of circulating microRNA in saliva." *Methods Mol Biol* 1024:183-90. doi: 10.1007/978-1-62703-453-1_14.

Gallo, A., C. Baldini, L. Teos, M. Mosca, S. Bombardieri, and I. Alevizos. 2012. "Emerging trends in Sjogren's syndrome: basic and translational research." *Clin Exp Rheumatol* 30 (5):779-84.

Gallo, A., S. Vella, F. Tuzzolino, N. Cuscino, A. Cecchettini, F. Ferro, M. Mosca, I. Alevizos, S. Bombardieri, P. G. Conaldi, and C. Baldini. 2019. "MicroRNA-mediated regulation of mucin-type O-glycosylation pathway: a putative mechanism of salivary gland dysfunction in Sjogren syndrome." *J Rheumatol.* doi: 10.3899/jrheum.180549.

Gauna, A. E., Y. J. Park, G. Nayar, M. Onate, J. O. Jin, C. M. Stewart, Q. Yu, and S. Cha. 2015. "Dysregulated co-stimulatory molecule expression in a Sjogren's syndrome mouse model with potential implications by microRNA-146a." *Mol Immunol* 68 (2 Pt C):606-16. doi: 10.1016/j.molimm.2015.09.027.

Gourzi, V. C., E. K. Kapsogeorgou, N. C. Kyriakidis, and A. G. Tzioufas. 2015. "Study of microRNAs (miRNAs) that are predicted to target the autoantigens Ro/SSA and La/SSB in primary Sjogren's Syndrome." *Clin Exp Immunol* 182 (1):14-22. doi: 10.1111/cei.12664.

Han, N., H. Lu, Z. Zhang, M. Ruan, W. Yang, and C. Zhang. 2018. "Comprehensive and in-depth analysis of microRNA and mRNA expression profile in salivary adenoid cystic carcinoma." *Gene* 678:349-360. doi: 10.1016/j.gene.2018.08.023.

Hayashi, T., and M. P. Hoffman. 2017. "Exosomal microRNA communication between tissues during organogenesis." *RNA Biol* 14 (12):1683-1689. doi: 10.1080/15476286.2017.1361098.

Hayashi, T., N. Koyama, Y. Azuma, and M. Kashimata. 2011. "Mesenchymal miR-21 regulates branching morphogenesis in murine submandibular gland in vitro." *Dev Biol* 352 (2):299-307. doi: 10.1016/j.ydbio.2011.01.030.

Hayashi, T., N. Koyama, E. W. Gresik, and M. Kashimata. 2009. "Detection of EGF-dependent microRNAs of the fetal mouse submandibular gland at embryonic day 13." *J Med Invest* 56 Suppl:250-2.

Hayashi, T., I. M. Lombaert, B. R. Hauser, V. N. Patel, and M. P. Hoffman. 2017. "Exosomal microRNA transport from salivary mesenchyme regulates epithelial progenitor expansion during organogenesis." *Dev Cell* 40 (1):95-103. doi: 10.1016/j.devcel.2016.12.001.

He, Q., X. Zhou, S. Li, Y. Jin, Z. Chen, D. Chen, Y. Cai, Z. Liu, T. Zhao, and A. Wang. 2013. "MicroRNA-181a suppresses salivary adenoid cystic carcinoma metastasis by targeting MAPK-Snai2 pathway." *Biochim Biophys Acta* 1830 (11):5258-66. doi: 10.1016/j.bbagen.2013.07.028.

Jevnaker, A. M., and H. Osmundsen. 2008. "MicroRNA expression profiling of the developing murine molar tooth germ and the developing murine submandibular salivary gland." *Arch Oral Biol* 53 (7):629-45. doi: 10.1016/j.archoralbio.2008.01.014.

Jiang, L. H., M. H. Ge, X. X. Hou, J. Cao, S. S. Hu, X. X. Lu, J. Han, Y. C. Wu, X. Liu, X. Zhu, L. L. Hong, P. Li, and Z. Q. Ling. 2015. "miR-21 regulates tumor progression through the miR-21-PDCD4-Stat3 pathway in human salivary adenoid cystic carcinoma." *Lab Invest* 95 (12):1398-408. doi: 10.1038/labinvest.2015.105.

Kapsogeorgou, E. K., V. C. Gourzi, M. N. Manoussakis, H. M. Moutsopoulos, and A. G. Tzioufas. 2011. "Cellular microRNAs (miRNAs) and Sjogren's syndrome: candidate regulators of autoimmune response and autoantigen expression." *J Autoimmun* 37 (2):129-35. doi: 10.1016/j.jaut.2011.05.003.

Kashimata, M., and T. Hayashi. 2018. "Regulatory mechanisms of branching morphogenesis in mouse submandibular gland rudiments." *Jpn Dent Sci Rev* 54 (1):2-7. doi: 10.1016/j.jdsr.2017.06.002.

Kiss, O., A. M. Tokes, S. Spisak, A. Szilagyi, N. Lippai, B. Szekely, A. M. Szasz, and J. Kulka. 2015. "Breast- and salivary gland-derived adenoid cystic carcinomas: potential post-transcriptional divergencies. A pilot study based on miRNA expression profiling of four cases and review of the potential relevance of the findings." *Pathol Oncol Res* 21 (1):29-44. doi: 10.1007/s12253-014-9770-1.

Kiss, O., A. M. Tokes, S. Vranic, Z. Gatalica, L. Vass, N. Udvarhelyi, A. M. Szasz, and J. Kulka. 2015. "Expression of miRNAs in adenoid cystic carcinomas of the breast and salivary glands." *Virchows Arch* 467 (5):551-62. doi: 10.1007/s00428-015-1827-3.

Konsta, O. D., Y. Thabet, C. Le Dantec, W. H. Brooks, A. G. Tzioufas, J. O. Pers, and Y. Renaudineau. 2014. "The contribution of epigenetics in Sjogren's Syndrome." *Front Genet* 5:71. doi: 10.3389/fgene.2014.00071.

Liang, Y., J. Ye, J. Jiao, J. Zhang, Y. Lu, L. Zhang, D. Wan, L. Duan, Y. Wu, and B. Zhang. 2017. "Down-regulation of miR-125a-5p is associated with salivary adenoid cystic carcinoma progression via targeting p38/JNK/ERK signal pathway." *Am J Transl Res* 9 (3):1101-1113.

Liu, L., Y. Hu, J. Fu, X. Yang, and Z. Zhang. 2013. "MicroRNA155 in the growth and invasion of salivary adenoid cystic carcinoma." *J Oral Pathol Med* 42 (2):140-7. doi: 10.1111/j.1600-0714.2012.01189.x.

Liu, X. Y., Z. J. Liu, H. He, C. Zhang, and Y. L. Wang. 2015. "MicroRNA-101-3p suppresses cell proliferation, invasion and enhances chemotherapeutic sensitivity in salivary gland adenoid cystic carcinoma by targeting Pim-1." *Am J Cancer Res* 5 (10):3015-29.

Lu, H., N. Han, W. Xu, Y. Zhu, L. Liu, S. Liu, and W. Yang. 2019. "Screening and bioinformatics analysis of mRNA, long non-coding RNA and circular RNA expression profiles in mucoepidermoid carcinoma of salivary gland." *Biochem Biophys Res Commun* 508 (1):66-71. doi: 10.1016/j.bbrc.2018.11.102.

Martini, D., A. Gallo, S. Vella, F. Sernissi, A. Cecchettini, N. Luciano, E. Polizzi, P. G. Conaldi, M. Mosca, and C. Baldini. 2017. "Cystatin S-a candidate biomarker for severity of submandibular gland involvement in Sjogren's syndrome." *Rheumatology (Oxford)* 56 (6):1031-1038. doi: 10.1093/rheumatology/kew501.

Matse, J. H., J. Yoshizawa, X. Wang, D. Elashoff, J. G. Bolscher, E. C. Veerman, E. Bloemena, and D. T. Wong. 2013. "Discovery and prevalidation of salivary extracellular microRNA biomarkers panel for the noninvasive detection of benign and malignant parotid gland tumors." *Clin Cancer Res* 19 (11):3032-8. doi: 10.1158/1078-0432.Ccr-12-3505.

Metzler, M. A., S. Appana, G. N. Brock, and D. S. Darling. 2015. "Use of multiple time points to model parotid differentiation." *Genom Data* 5:82-8. doi: 10.1016/j.gdata.2015.05.005.

Metzler, M. A., S. G. Venkatesh, J. Lakshmanan, A. L. Carenbauer, S. M. Perez, S. A. Andres, S. Appana, G. N. Brock, J. L. Wittliff, and D. S. Darling. 2015. "A systems biology approach identifies a regulatory network in parotid acinar cell terminal differentiation." *PLoS One* 10 (4):e0125153. doi: 10.1371/journal.pone.0125153.

Michael, A., S. D. Bajracharya, P. S. Yuen, H. Zhou, R. A. Star, G. G. Illei, and I. Alevizos. 2010. "Exosomes from human saliva as a source of microRNA biomarkers." *Oral Dis* 16 (1):34-8. doi: 10.1111/j.1601-0825.2009.01604.x.

Mitani, Y., D. B. Roberts, H. Fatani, R. S. Weber, M. S. Kies, S. M. Lippman, and A. K. El-Naggar. 2013. "MicroRNA profiling of salivary adenoid cystic carcinoma: association of miR-17-92 upregulation with poor outcome." *PLoS One* 8 (6):e66778. doi: 10.1371/journal.pone.0066778.

Nelson, C., V. Ambros, and E. H. Baehrecke. 2014. "miR-14 regulates autophagy during developmental cell death by targeting ip3-kinase 2." *Mol Cell* 56 (3):376-88. doi: 10.1016/j.molcel.2014.09.011.

Pauley, K. M., C. M. Stewart, A. E. Gauna, L. C. Dupre, R. Kuklani, A. L. Chan, B. A. Pauley, W. H. Reeves, E. K. Chan, and S. Cha. 2011. "Altered miR-146a expression in Sjogren's syndrome and its functional role in innate immunity." *Eur J Immunol* 41 (7):2029-39. doi: 10.1002/eji.201040757.

Ramachandran, I., V. Ganapathy, E. Gillies, I. Fonseca, S. M. Sureban, C. W. Houchen, A. Reis, and L. Queimado. 2014. "Wnt inhibitory factor 1 suppresses cancer stemness and induces cellular senescence." *Cell Death Dis* 5:e1246. doi: 10.1038/cddis.2014.219.

Rebustini, I. T., T. Hayashi, A. D. Reynolds, M. L. Dillard, E. M. Carpenter, and M. P. Hoffman. 2012. "miR-200c regulates FGFR-dependent epithelial proliferation via Vldlr during submandibular gland branching morphogenesis." *Development* 139 (1):191-202. doi: 10.1242/dev.070151.

Santos, P. R. B., C. M. Coutinho-Camillo, F. A. Soares, V. S. Freitas, D. S. Vilas-Boas, F. C. A. Xavier, C. A. G. Rocha, I. B. de Araujo, and J. N. Dos Santos. 2017. "MicroRNAs expression pattern related to mast cell activation and angiogenesis in paraffin-embedded salivary gland tumors." *Pathol Res Pract* 213 (12):1470-1476. doi: 10.1016/j.prp.2017.10.012.

Shi, H., N. Cao, Y. Pu, L. Xie, L. Zheng, and C. Yu. 2016. "Long non-coding RNA expression profile in minor salivary gland of primary Sjogren's syndrome." *Arthritis Res Ther* 18 (1):109. doi: 10.1186/s13075-016-1005-2.

Simpson, R. H., A. Skálová, S. Di Palma, and I. Leivo. 2014. "Recent advances in the diagnostic pathology of salivary carcinomas." *Virchows Arch* 465 (4):371-84. doi: 10.1007/s00428-014-1639-x.

Sun, L., B. Liu, Z. Lin, Y. Yao, Y. Chen, Y. Li, J. Chen, D. Yu, Z. Tang, B. Wang, S. Zeng, S. Fan, Y. Wang, Y. Li, E. Song, and J. Li. 2015. "MiR-320a acts as a prognostic factor and Inhibits metastasis of salivary adenoid cystic carcinoma by targeting ITGB3." *Mol Cancer* 14:96. doi: 10.1186/s12943-015-0344-y.

Talotta, R., V. Mercurio, S. Bongiovanni, C. Vittori, L. Boccassini, F. Rigamonti, A. Batticciotto, F. Atzeni, D. Trabattoni, P. Sarzi-Puttini, and M. Biasin. 2019. "Evaluation of salivary and plasma microRNA expression in patients with Sjogren's syndrome, and correlations with clinical and ultrasonographic outcomes." *Clin Exp Rheumatol*.

Tandon, M., A. Gallo, S. I. Jang, G. G. Illei, and I. Alevizos. 2012. "Deep sequencing of short RNAs reveals novel microRNAs in minor salivary glands of patients with Sjogren's syndrome." *Oral Dis* 18 (2):127-31. doi: 10.1111/j.1601-0825.2011.01849.x.

Veit, J. A., K. Scheckenbach, P. J. Schuler, S. Laban, P. S. Wiggenhauser, J. Thierauf, J. P. Klussmann, and T. K. Hoffmann. 2015. "MicroRNA expression in differentially metastasizing tumors of the head and neck: adenoid cystic versus squamous cell carcinoma." *Anticancer Res* 35 (3):1271-7.

von Ahsen, I., R. Nimzyk, M. Klemke, and J. Bullerdiek. 2008. "A microRNA encoded in a highly conserved part of the mammalian

HMGA2 gene." *Cancer Genet Cytogenet* 187 (1):43-4. doi: 10.1016/j.cancergencyto.2008.07.009.

Wang, C., T. Li, F. Yan, W. Cai, J. Zheng, X. Jiang, and J. Sun. 2018. "Effect of simvastatin and microRNA-21 inhibitor on metastasis and progression of human salivary adenoid cystic carcinoma." *Biomed Pharmacother* 105:1054-1061. doi: 10.1016/j.biopha.2018.05.157.

Wang, K. C., and H. Y. Chang. 2011. "Molecular mechanisms of long noncoding RNAs." *Mol Cell* 43 (6):904-14. doi: 10.1016/j.molcel.2011.08.018.

Wang, S., L. Zhang, P. Shi, Y. Zhang, H. Zhou, and X. Cao. 2018. "Genome-wide profiles of metastasis-associated mRNAs and microRNAs in salivary adenoid cystic carcinoma." *Biochem Biophys Res Commun* 500 (3):632-638. doi: 10.1016/j.bbrc.2018.04.122.

Wang, W. W., B. Chen, C. B. Lei, G. X. Liu, Y. G. Wang, C. Yi, Y. Y. Wang, and S. Y. Zhang. 2017. "miR-582-5p inhibits invasion and migration of salivary adenoid cystic carcinoma cells by targeting FOXC1." *Jpn J Clin Oncol* 47 (8):690-698. doi: 10.1093/jjco/hyx073.

Wang, Y., G. Zhang, L. Zhang, M. Zhao, and H. Huang. 2018. "Decreased microRNA-181a and -16 expression levels in the labial salivary glands of Sjogren syndrome patients." *Exp Ther Med* 15 (1):426-432. doi: 10.3892/etm.2017.5407.

Xin, M., H. Liang, H. Wang, D. Wen, L. Wang, L. Zhao, M. Sun, and J. Wang. 2019. "Mirt2 functions in synergy with miR-377 to participate in inflammatory pathophysiology of Sjogren's syndrome." *Artif Cells Nanomed Biotechnol* 47 (1):2473-2480. doi: 10.1080/21691401.2019.1626413.

Yan, F., C. Wang, T. Li, W. Cai, and J. Sun. 2018. "Role of miR-21 in the growth and metastasis of human salivary adenoid cystic carcinoma." *Mol Med Rep* 17 (3):4237-4244. doi: 10.3892/mmr.2018.8381.

Yan, T., J. Shen, J. Chen, M. Zhao, H. Guo, and Y. Wang. 2019. "Differential expression of miR-17-92 cluster among varying histological stages of minor salivary gland in patients with primary Sjogren's syndrome." *Clin Exp Rheumatol.*

Yang, Y., L. Peng, W. Ma, F. Yi, Z. Zhang, H. Chen, Y. Guo, L. Wang, L. D. Zhao, W. Zheng, J. Li, F. Zhang, and Q. Du. 2016. "Autoantigen-targeting microRNAs in Sjogren's syndrome." *Clin Rheumatol* 35 (4):911-7. doi: 10.1007/s10067-016-3203-3.

Ying, S. Y., D. C. Chang, and S. L. Lin. 2008. "The microRNA (miRNA): overview of the RNA genes that modulate gene function." *Mol Biotechnol* 38 (3):257-68. doi: 10.1007/s12033-007-9013-8.

Zhan, C., X. Yang, X. Yin, and J. Hou. 2019. "Exosomes and other extracellular vesicles in oral and salivary gland cancers." *Oral Dis*. doi: 10.1111/odi.13172.

Zhang, X., M. Cairns, B. Rose, C. O'Brien, K. Shannon, J. Clark, J. Gamble, and N. Tran. 2009. "Alterations in miRNA processing and expression in pleomorphic adenomas of the salivary gland." *Int J Cancer* 124 (12):2855-63. doi: 10.1002/ijc.24298.

In: Salivary Glands
Editor: Alicia S. Bryant

ISBN: 978-1-53617-497-7
© 2020 Nova Science Publishers, Inc.

Chapter 2

PAROTID GLAND NEOPLASM: DIAGNOSIS AND MANAGEMENT

Salma M. Al Sheibani[*], *MD*
ENT department, Al Nahdha Hospital, Muscat, Sultanate of Oman

ABSTRACT

The parotid gland is the largest of the major salivary glands and the most common site for salivary tumors, the majority of which are benign. Parotid gland tumors encompass a mixture of different histologies that can be categorized into low, moderate, and high risk classifications based on their unique biologies and clinical outcomes. Fine needle aspiration should be considered as a part of the diagnostic assessment, but it has varying sensitivities and specificities. Anatomic imaging of the parotid glands, such as ultrasound, CT scans, and MRI, is crucial for planning of surgical management.

Surgery is the primary modality for management of these tumors. Facial nerve monitoring can be selectively used to decrease the morbidity associated with this type of surgery. The complications of parotid surgery can be classified into intra-operative and post-operative. The latter type can be subdivided into early and late complications. This operation continues

---

[*] Corresponding Author Email: salsheibani@gmail.com.

to be a challenge on account of the wide range of tumors found and the differences in size and location. It is, therefore, mandatory that an experienced surgeon performs this operation.

Low risk salivary gland tumors are often treated with adequate surgical resection and do not require adjuvant radiation because they display very low rates of recurrence. Adjuvant radiation therapy is required in moderate risk salivary gland tumors as local recurrence can occur frequently even after optimal surgical management. High risk salivary gland tumors carry a substantial risk of spread to the ipsilateral neck nodes and show an improvement in local control and survival with adjuvant radiation therapy. Progress in modern radiation techniques now permit treatment of salivary gland tumors with fewer adverse effects and excellent local and regional control. Chemotherapy remains to have a palliative role in the management of salivary gland tumors, though researchers in this field are trying to identify a therapeutic role for chemotherapy in order to improve overall survival.

This chapter reviews parotid tumors, diagnostic assessment, surgical approaches, use of neck dissection, and use of radiation and chemotherapy. In addition, the management of the facial nerve in malignant parotid tumors is highlighted. Furthermore, the most common complications of parotid surgery and relative management are taken into consideration.

Keywords: parotid tumor, salivary gland tumor, parotid cancer, parotidectomy, facial nerve, radiotherapy, chemotherapy, Parotid neoplasms, targeted therapy, FNAC

INTRODUCTION

The parotid gland is the largest major salivary gland and has the highest incidence, approximately 70-80%, of overall benign salivary gland tumors. Pleomorphic adenoma constitutes 40 -70% of all benign tumors followed by Warthin's tumor in 25-30% [1-5].

The incidence of salivary gland malignancies is rare and the reported frequency varies among different countries: acinic cell carcinoma (10–18%), adenoid cystic carcinoma (9–15%), adenocarcinoma (13–32%), mucoepidermoid carcinoma (11–31%) carcinoma ex pleomorphic adenoma (7–13%), squamous cell carcinoma (9–17%) and salivary duct carcinoma (3–6%) [4-7].

Table 1. WHO histological classification of salivary gland tumors 2017 [10, 11]

Malignant tumors	Basal cell adenoma 8147/0
Mucoepidermoid carcinoma 8430/0	Warthin tumor 8561/0
Acinic cell carcinoma 8550/3	
Secretory carcinoma 8502/3	
Adenoid cystic carcinoma 8200/3	Oncocytoma 8290/0
Polymorphous adenocarcinoma 8525/3	Lymphoadenoma 8563/0
Epithelial – myoepithelial carcinoma 8562/3	Cystadenoma 8440/0
Clear cell carcinoma 8310/3	Sialadenoma papilliferum 8406/0
Basal cell adenocarcinoma 8147/3	Ductal papillomas 8503/0
Sebaceous adenocarcinoma 8410/3	Sebaceous edenoma 8410/0
Intraductal carcinoma 8500/2	Canalicualr adenoma and other ductal adenomas 8149/0
Cystadenocarcinoma 8440/3	**Other epithelial lesions**
Adenocarcinoma, NOS 8140/3	Sclerosing polycystic adenosis
Salivary duct carcinoma 8500/3	Nodular oncolcytic hyperplasia
Myoepithelial carcinoma 8982/3	Lymphepithelial lesions
Carcinoma ex pleomorphic adenoma 8941/3	Intercalated duct hyperplasia
Carcinosarcoma 8980/3	**Soft tissue lesions**
Poorly differentiated carcinoma	Haemagioma 9120/0
Neuroendocrine and non-neuroendocrine	Lipoma / sialolipma 8850/0
Undifferentiated carcinoma 8020/3	Nodular fasciitis 8828/0
Large cell neuroendocrine carcinoma 8013/3	**Haematolymphoid tumors**
Small cell neuroendocrine carcinoma 8041/3	Extranodal marginal zone lymphoma of MALT 9699/3
Lymphoepithelial carcinoma 8082/3	The morphology codes are from International Classification of Disease for Oncology (ICD-0) (742A). Behavior is coded /0 for benign tumors, /1 for unspecified, borderline or uncertain behavior, /2 for carcinoma in situ and grade III intraepithelial neoplasia, and /3 for malignant tumours.
Squamous cell carcinoma 8070/3	
Oncocystic carcinoma 8290/3	
Borderline tumor	The classification is modified from the previous WHO classification, taking into account changes in our understanding of these lesions.
Sialoblastoma	*These new codes were approved by the IARC/WHO Committee for ICD-0.
Benign tumors	*Italic*: Provisional tumour entities ** Grading according to the 2013 WHO classification of tumours of soft tissue and bone.
Pleomorphic adenoma 8940/0	
Myoepithelial 8982/0	

Although salivary gland tumors are uncommon in children, 20-30% of them are malignant which are usually low-grade mucoepidermoid carcinomas [8].

Salivary gland tumors are rare and have wide histology diversity and clinical manifestations. The World Health Organization (WHO) classification has been modified multiple times, the last being in 2017 (Table 1) [9-11]. In this edition, new entities have been added such as mammary analogue secretory carcinoma that earlier used to be included within acinic cell carcinoma. In addition, new subsections of other epithelial lesions were also added that includes the addition of hyperplastic lesions such as sclerosing polycystic adenosis, nodular oncocytic hyperplasia, and intercalated duct hyperplasia. Many tumors have been compressed into their wider categories or reclassified including adenocarcinoma subtypes that were placed under adenocarcinoma NOS. Moreover,-specific tumor grades have been removed. This is specifically related to polymorphous adenocarcinoma, in which low grade had been removed as it contains a cribriform subtype, which has an aggressive behavior. Carcinoma ex. pleomorphic adenoma that arise within the pleomorphic adenoma has been classified into intracapsular, minimally invasive, and widely invasive lesions. Some tumor entities have been collapsed into their parent entities; the exophytic and inverted ductal papillomas were merged under ductal papilloma. Similarly, sebaceous and non-sebaceous lymphadenomas were joined under lymphadenoma. Furthermore, a streamlined grading system has been achieved for mucoepidermoid carcinoma and adenoid cystic carcinoma.

In soft tissue lesions, in addition to hemangioma, lipoma and nodular fasciitis were added. The term lymphoproliferative disorders was restricted to those frequently seen in salivary glands with an emphasis on extranodal marginal zone and mucosa-associated lymphoma (MALTsyndrome) as the most common type of lymphomas in this area [9, 11, 12].

DIAGNOSTIC ASSESSMENT

The diagnosis and management of salivary gland tumors is dependent on the clinical presentation, imaging, and cytology and/or histology results. The majority of parotid tumors present with a painless, palpable lump. Presences of pain, facial palsy, and rapid growth or fixation to underlying or overlying structures are indicative of malignancy. These tumors are mostly located in the superficial lobe and can be seen by ultrasound [2, 5].

Ultrasound can be used to assist in discriminating benign from malignant tumors in the majority of cases and provides information about the status of tumor and neck node. Although it is an operator dependent procedure, it is safe with a high diagnostic accuracy when combined with FNAC [13, 14]. CT and MRI can be used to determine the exact tumor location, whether it's in superficial or deep lobe (Figure 1, 2, 3 and 4) [2]. CT scan is preferable when there is a suspicion of skull base or mandible involvement, in addition to assessment of lymph nodes involvement. MRI is superior in soft tissue delineation and in outlining the interface between tumor and normal salivary gland (Figure 5).

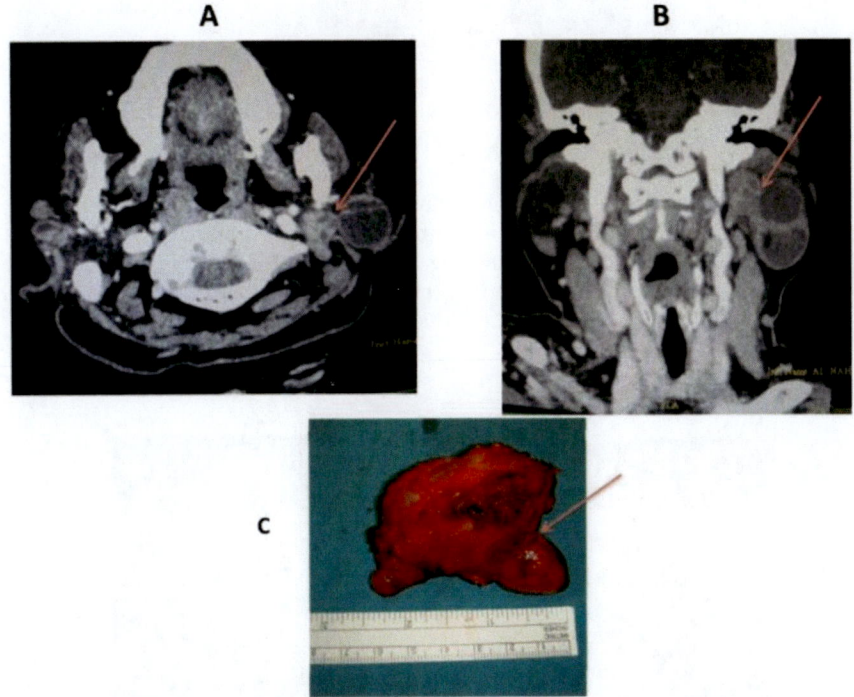

Figure 1. A, B: An axial and coronal CT scan of a 50 year old lady demonstrating a dumbbell tumor extending from superficial to deep lobe of left parotid gland with a constriction at the stylomandibular tunnel (red arrow). C: The excised tumor showing the two components with the constriction (red arrow).

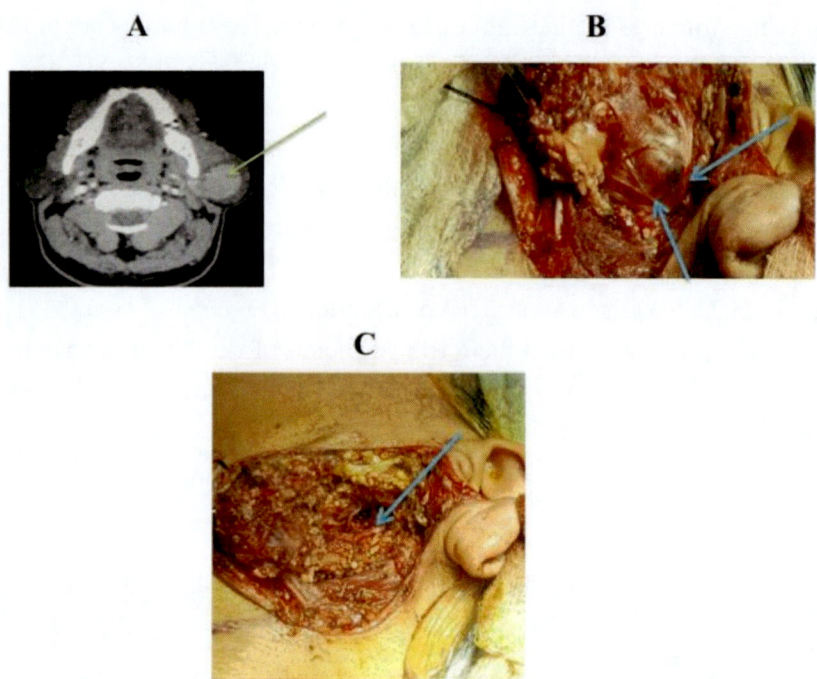

Figure 2. A. An axial CT scan with intravenous contrast demonstrating a left deep lobe tumor that is heterogeneously enhancing with contrast (green arrow). B- An intraoperative image demonstrating how the deep lobe parotid tumor pushing the facial nerve and branches over the tumor surface (blue arrows). C- Surgical bed after tumor excision. Facial nerve trunk and branches (blue arrow).

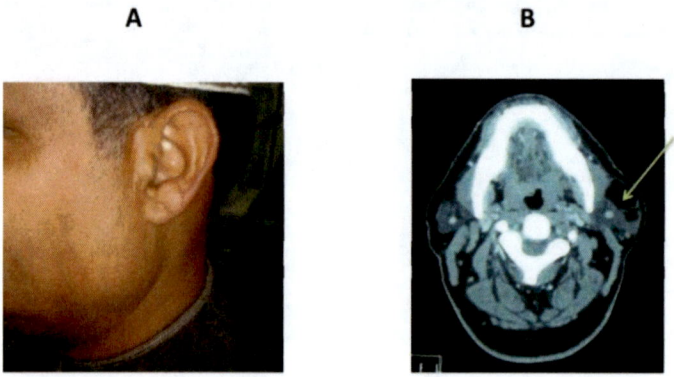

Figure 3. A- A 45- year old gentleman presented with left parotid mass. B- An Axial CT scan image demonstrating and a hypodense mass in the superficial lobe (green arrow) suggestive of lipoma.

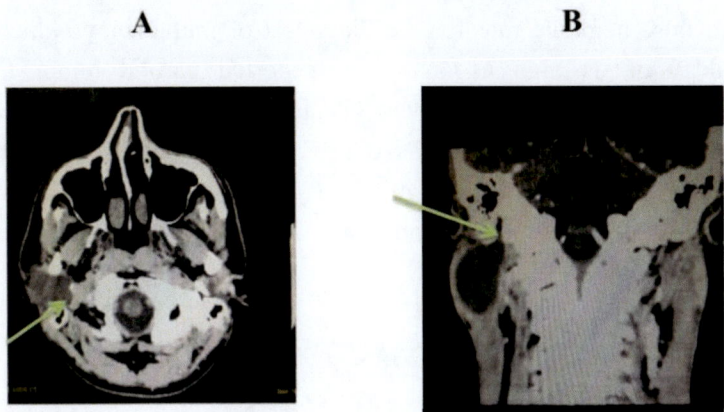

Figure 4. A and B. An axial and coronal CT scan of a right parotid tumor extending into the mastoid foramen. The mastoid foramen is widened on right side compared to left side (green arrow). Whenever such finding encountered MRI should be requested, as this finding is highly suggestive neurogenic tumor such as facial nerve neuroma or schwanoma.

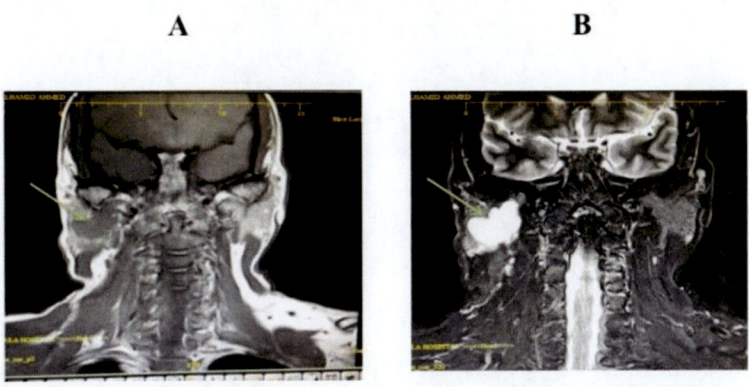

Figure 5. A- A coronal T1- weighted MRI images demonstrating a hypointense appearance of pleomorphic adenoma of right parotid gland. B- The tumor is hyperintense in T2- weighted MRI.

The role of FNAC in parotid neoplasms management remains controversial. Most of the surgeons do not rely on FNAC results alone and favor intraoperative frozen section for procedure modification and resection extent [15-17].

The reported rate of parotid FNAC sensitivity ranged from 54 to 92% and specificity from 86 to 100% [18-30]. This low sensitivity is attributed to the high false-negative rate for the diagnosis of malignancy, which was reported to be as high as 46%. The reported parotid tumor FNAC histological typing ranged from 29 to 84% [17, 31].

O'Brien suggested doing FNAC in order to justify conservative management in those with a benign tumor. For example, for those patients with poor general condition and were not fit for surgical resection (Figure 6) [32].

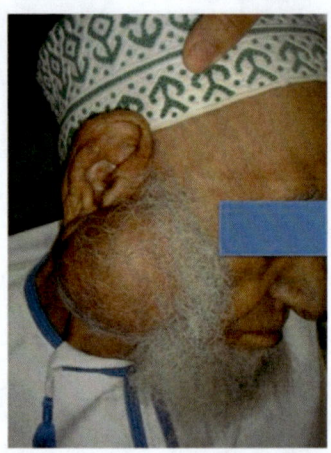

Figure 6. A 100-year-old gentleman with multiple comorbidities and poor general condition presented with right parotid mass. FNAC came as pleomorphic adenoma. He is not fit for surgery and since the FNAC is benign he was kept under observation.

Intraoperative frozen section of a partial parotidectomy specimen may help to determine the presence or absence of malignancy and guide the surgeon for the extent of surgical resection rather than a completion parotidectomy that carries a significant risk for facial nerve function.

STAGING

Parotid gland malignancy are classified according to the 8th edition of the Union International Contra Cancer/American Joint Committee (UICC/AJCC) cancer staging system as shown in Table 2 [36].

Table 2. TNM Classification of parotid cancer [36]

T- Primary tumor (identical for cT and pT)	pN-Regional Lymph Nodes
TX: Primary tumor cannot be assessed	pNX: Regional lymph nodes cannot be assessed
T0: No evidence of primary tumor	pN0: No regional node metastasis
T1: Tumor 2 cm or less in greatest dimension without extraparenchymal extension *	pN1: Metastasis in a single ipsilateral lymph node, 3 cm or less in greatest dimension without extranodal extension (ENE-)
T2: Tumor more than 2 cm but not more than 4 cm in greatest dimension without extraparenchymal extension *	pN2a: Metastasis in a single ipsilateral lymph node, 3 cm or less in greatest dimension with extranodal extension (ENE+) or
T3: Tumor more than 4 cm and/or tumor with extraparenchymal extension *	metastasis in a single ipsilateral lymph node, more than 3 cm but not more than 6 cm in greatest dimension without extranodal extension (ENE-)
T4a: Tumor invades skin, mandible, ear canal and/or facial nerve	
T4b: Tumor invades base of skull, and/or pterygoid plates, and/ or encases caroid artery	pN2b: Metastasis in multiple ipsilateral lymph node, none more than 6 cm in greatest dimension, without extranodal extension (ENE-)
Note: * Extraparenchymal extension is clinical or macroscopic evidence of invasion of soft tissues or nerve, except those listed under T4a and 4b. Microscopic evidence alone does not constitute extraparenchymalextension for classification purposes.	pN2c: Metastasis in bilateral or contralateral lymph nodes, none more than 6 cm in greatest dimension, without extranodal extension (ENE-)
cN- Regional Lymph Nodes	pN3a: Metastasis in a lymph node more than 6 cm in greatest dimension without extranodal extension (ENE-)
N1: Metastasis in a single ipsilateral lymph node, 3 cm or less in greatest dimension without extranodal extension (ENE-).	
N2a: Metastasis in a single ipsilateral lymph node, more than 3 cm but more than 6 cm in greatest dimension without extranodal extension (ENE-)	PN3b Metastasis in a lymph node more than 3 cm in greatest dimension with extranodal extension (ENE+) or multiple ipsilateral, or any contralateral, or bilateral node (s) with extranodal extension (ENE+)
N2b: Metastasis in multiple ipsilateral lymph nodes, none more than 6 cm in greatest dimension, without extranodal extension (ENE-)	**Stage Grouping**
N2c: Metastasis in bilateral or contralateral nodes, none more than 6 cm in greatest dimension, without extranodal extension (ENE-)	Stage 0: Tis N0 M0
	Stage I: T1 N0 M0
	Stage II: T2 N0 M0
N3a: Metastasis in a lymph node more than 6 cm in greatest dimension without extranodal extension (ENE-)	Stage III: T3 N0 M0 T1,T2,T3 N1 M0
N3b: Metastasis in a single or multiple lymph nodes with clinical extranodal extranodal extension (ENE+)**	Stage IVA: T1,T2,T3 N2 M0 T4a N0,N1,N2 M0
	Stage IVB: T4a Any N M0 Any T N3 M0
	Stage IVC: Any T Any N M1

Notes: Midline nodes are considered ipsilateral nodes.

Parotid Gland Tumors

There are a diverse group of benign and malignant parotid tumors. The section below reviews the most common and peculiar parotid tumors.

Benign Tumors

Pleomorphic Adenoma
PA is the most common salivary gland tumor in all ages, which is most commonly seen in the fourth to sixth decades. The mean age at the presentation is around 45 years. In old age, it is more commonly seen in male compared to female patients. Its incidence has been reported to rise 15-20 years after exposure to radiation [11, 37, 38, 39].

Around 90% of pleomorphic adenomas are located in the superficial lobe, and only 10% are found in deep lobe. It is a mixed benign tumor in which the parenchyma consisted of a variable epithelial and myoepithelial cells embedded in a stroma of mucoid, myxoid, chondroid or osteoid origin. The ratio of parenchyma to stroma is used to classify these tumors into parenchyma-rich and stroma-rich variants [11, 40-43]. The capsule is the most important structure of pleomorphic adenoma and there is no agreement whether it is a true capsule or it is formed by the tumor compression of the pre-existing connective tissue.

It was found that it's thick in parenchyma-rich tumors compared to stroma-rich tumors. It was reported that up to two thirds of stroma-rich tumors have focal absence of the capsule that leads to pseudopodias or tumor satellites, which subsequently resulted into a multifocal pleomorphic adenoma [39, 44-47].

The recurrence rate is low in a completely resected tumor. The malignant transformation has been reported up to 6.2% and it's more commonly seen in patients with multiple recurrence, male sex, old age and in tumors located in the deep parotid lobe [11]. The most commonly transformed malignancy is carcinoma ex. pleomorphic adenoma. Pleomorphic adenoma should be treated by surgical resection.

Warthin's Tumor

Warthin's tumor is the second most common benign parotid gland, described in 1929 by Aldred Warthin. It's more common in male patients, with a male to female ratio of 4:1. It affect patients in their sixth and seventh decades of life [48, 49]. It is also known as papillary cystadenoma lymphatosum, cystadenolymphoma, and adenolymphoma [11]. It has been linked to smoking. Its incidence among females increased which could be explained by increased incidence of smoking habit among women [50]. It's multifocal in 6-20% and bilateral in 5-15% [11, 51-54]. It is commonly seen in the superficial lobe, especially the parotid tail. It appears as a hot lesion on technetium-99m pertechnetate imaging [11].

It consisted of variable amounts of papillary-cystic structures lined by oncocytic epithelial cells and a lymphoid stroma [11].

Warthin's tumor is bilateral in around 10%. It should be treated by complete surgical excision. The recurrence rate ranged from 0-13%, which is due to the multifocal nature and is dependent on the extent of surgical resection [11, 48, 49, 53]. Malignant transformation has been reported in only 0.1% of cases and arises from both epithelial and lymphoid component of the tumor [11, 49, 51].

Metastasizing Pleomorphic Adenoma

Metastasizing Pleomorphic Adenoma is similar to pleomorphic adenoma histologically, but it behaves in a malignant manner and is potentially fatal. It was kept as a subsection under benign, pleomorphic adenoma in the 4th WHO classification of head and neck tumors, however it has aggressive behavior [11]. It appears after multiple recurrences and has a tendency to metastasize to regional lymph nodes and distant metastasis. It most frequently metastasizes to bones and the lungs [12, 55, 56]. The proposed theory for its development is through haematogenous route by vascular implantation at the time of the initial surgery.

The reported latency period between the initial presentation and the appearance of recurrence and distant metastasis ranged from 7 to 16 years [55, 57, 58].

The mortality rate is 50% at 5 years [59]. 40% of the patients may die with the disease, 47% are likely to live free of disease, and 13% will live with it [55, 60].

Metastasizing Pleomorphic Adenoma should be treated with total parotidectomy and facial nerve preservation if it's not involved followed by adjuvant radiotherapy [61].

Sclerosing Polycystic Adenosis

Sclerosing polycystic adenosis is recently described and it has similar characteristics and appearance to breast sclerosing adenosis and adenosis tumors. It affects major and minor salivary glands, but is mostly seen in the parotid gland. It has a wide age range, though it is mostly seen in 4th decade. It has a slight female predominance. Sometimes it's multifocal. The greatest dimension of tumors size ranged from a few millimeters to 7 cm [12, 62].

The etiology is unknown, however it may be associated with Epstein Barr virus and some studies suggested it be put under neoplastic categories as it's a colonal process [63]. The recurrence rate reached up to 11%, which is due to multifocality and incomplete resection. The ductal component of sclerosing polycystic adenosis may proliferate and show similar characteristics to intraductal carcinoma. Nevertheless, to date only one case of a salivary ductal carcinoma has been reported in a sclerosing polycystic adenosis that has multiple recurrences [64].

Malignant Tumors

Mucoepidermoid Carcinoma

Mucoepidermoid carcinoma is the most common salivary gland malignancies and mostly affect the parotid gland. It constitutes 30 to 40% of all salivary gland malignancies and up to 50% of parotid malignancies [65, 66]. There is no grading system recognized in the 4th WHO edition. Only the general features of low, intermediate, and high grade tumors are defined [11]. It is more commonly seen in children and young adults. It may develop secondary to radiation or chemotherapy during childhood [11, 66].

Mucoepidermoid carcinoma consists of mucinous, intermediate (clear), and epidermoid cells forming solid and cystic components. The low grade mucoepidermoid carcinoma contains more mucinous cells compared to the other cells. The high grade mucoepidermoid carcinoma consist of more epidermoid cells compared to the other cells. The intermediate grade mucoepidermoid carcinoma contains more mucinous and intermediate cells.

The majority of mucoepidermoid carcinoma have t(11; 19) (q 21; p13) translocation and CRTC1-MAML2 gene fusion while a small group exhibited (11; 15) (q21;q26) translocation and CRTC3-MAML2 gene fusion. Tumors, which have translocation and gene fusion, tend to be of low and intermediate grades and is mostly seen in young patients [11].

High histological grade, large tumor size, nodal disease, metastatic disease, perineural invasion, and positive surgical margins are pathological features correlated with poor survival outcome. The prognosis is worsening from low to high grade mucoepidermoid carcinoma.

Surgery is the treatment of choice. Low and intermediate grade mucoepidermoid carcinoma are cured with surgery alone, however, high grade mucoepidermoid carcinoma is more aggressive and required adjuvant radiotherapy [11, 66].

Adjuvant radiotherapy is indicated for high grade mucoepidermoid carcinoma, stage III and IV tumors, tumors with extensive perineural or vascular invasion or with extraglandular extension, tumors of the deep lobe of the parotid, and incompletely excised tumors [72, 73].

The reported 5-year overall survival ranged from 68%-75% [69-71]. The 10 year overall survival rates for low, intermediate, and high grade MECs are around 90%, 70%, and 25% respectively [11].

Acinic Cell Carcinoma

Acinic cell carcinoma is an uncommon salivary gland malignancy that constitutes 6-7% of salivary gland neoplasms [11, 60, 73, 74]. More than 90-95% arise from the parotid gland. It is the second most common salivary gland malignancy. There is a slight female predominance with a female to male ratio of 1.5: 1. About 35% occur in patients aged more than 60 years and 4% are aged <20 years [11, 74].

The main presentation is a slowly growing mobile mass. However, some of the patients might present with a multi-nodular mass that might be fixed to skin with occasional pain. Facial nerve paralysis is seen 5-10% of the cases [11, 74].

The presence of cytoplasmic zymogen secretory granules is diagnostic of acinic cell carcinoma. A variety of histological growth forms can be seen such as solid, microcystic, follicular, and papillocystic. It has a multifocal origin and sometimes bilateral location. Acinic cell carcinoma is typically a low grade malignancy, however cases of acinic cell carcinoma with high grade transformation has been reported. These cases behave aggressively and frequently affect the facial nerve with high propensity to recur and metastasize to lymph nodes and the lungs with poor prognosis [11, 74, 75].

A 35% recurrence rate has been reported. The survival rate is 90% at 20 year [11].

The treatment is surgical resection. Adjuvant radiotherapy is given to the cases with high-grade transformation. The poor prognostic features are large T size, deep lobe involvement, and incomplete resection.

Adenoid Cystic Carcinoma

Adenoid cystic carcinoma is a malignant salivary gland tumor, which is slow-growing, characterized by local recurrence, late distant metastasis, and fatal outcome. It is an uncommon tumor and the reported incidence in the parotid gland ranged between 6-10% [76, 77]. It consisted of epithelial and myoepithelial neoplastic cells that form mixed patterns, comprising of tubular, cribriform and solid forms. The solid form is the most aggressive form with a tendency of lymph nodes metastasis [11, 78].

The most important adverse prognostic factor is the amount of existing solid component [80-83]. It might undergo high- grade transformation, which is an aggressive tumor with extraglandular extension and positive surgical margins with a high risk of lymph nodes involvement that is reported as high as 57.9% [6, 78, 82-86].

The risk of nodal metastasis in conventional adenoid cystic carcinoma has been reported in the rage from 5% to 25% [77].

Sometimes the histological distinction of adenoid cystic carcinoma high grade transformation from conventional adenoid cystic carcinoma is difficult. There are specific unique morphologic features of adenoid cystic carcinoma high grade transformation that can be detected by immunohistochemistry, such as the manifestation of micropapillae and squamoid areas, fibrocellular desmoplasia, and the loss of an abluminal layer of myoepithelial cells. Other features that raise the suspicion of high-grade transformation is the presence of strong p53 immunoreactivity in >50% of cells. In addition to, the other major criteria that was reported by Seethala et al. (2007) [78].

Adenoid cystic carcinoma high grade transformation is an aggressive variant of adenoid cystic carcinoma associated with even more aggressive behavior than the solid adenoid cystic carcinoma. High-grade transformation may occur in any of the 3 subtypes and it has a high tendency of lymph nodes metastasis in which elective neck dissection should be considered [78].

There are different histopathological grading systems described by Perzin et al. (1978) [80], Szanto et al. (1984) [81], Spiro et al. (1974) [88] and by Weert et al. (2015) [88]. All of these grading systems are in agreement that the presence and the amount of solid component present in the tumor are an indication of a more aggressive tumor and poor prognosis.

The treatment of choice is surgery followed by radiotherapy. The local recurrence of adenoid cystic carcinoma is rare, however it is characterized by distant metastasis in more than 50% of the cases. The most common site of distant metastasis is the lungs, followed by bone, liver, and brain [11].

The reported overall survival rates at 5 and 10 years were 55-90% and 30–70% respectively [79, 89]. The overall survival continues to drop after 5 years.

The prognosis is affected by advanced clinical stage, patient age, presence of perineural invasion, lymph nodes involvement, positive surgical margins, and solid pattern growth. Perineural invasion is a reliable prognostic factor for distant metastasis development. The presence of distant metastasis is indicative of poor prognosis [11, 90]. The incidence of distant metastasis has ranged from 25-55%, and only 20% of patients survive for 5-years [11, 90-92].

Adenocarcinoma, NOS

Adenocarcinoma, NOS, represents about 10-15% of all salivary gland carcinomas. It affects a wide age range with an average of 58 years. It's very rare in children. More than 50% appears in the parotid gland and about 40% occurred in minor salivary glands [11].

The most common presentation is a symptomatic solitary firm or cystic mass [8, 11].

Histologically, it consists of ductal or glandular proliferations with or without cystic formation. Based on the degree of cellular atypia, it can be graded as low, intermediate, or high grade. Ductal and glandular structures are seen frequently in low and intermediate grade tumors.

Histologically, it should be differentiated from salivary ductal carcinoma, high grade mucoepidermoid carcinoma, polymorphous adenocarcinoma, metastatic adenocarcinoma, acinic cell carcinoma, and from tumors that contain myoepitheial and basal cell component.

Mucinous and intestinal-type adenocarcinoma are aggressive and rare subtypes. The high grade adenocarcinoma is also an aggressive tumor and the incidence of distant metastasis is around 40%, which is directly affecting the survival [8]. The 15-year survival rates for low, intermediate, and high grade tumors reported as 54%, 31%, and 3%, respectively. The low and intermediate grade adenocarcinoma are treated by wide local excision. The high grade tumors should be treated by wide local resection with elective neck dissection and adjuvant radiotherapy [8].

Salivary Duct Carcinoma

Salivary ductal carcinoma is also known as high-grade ductal carcinoma.

It represents around 10% of salivary malignancy and mostly affects the parotid gland. It is most commonly seen in elderly male patients in the 6th or 7th decades. It's an aggressive epithelial malignancy, showing similar histological features of breast high grade mammary ductal carcinoma. It may arise de nova or as a constituent of carcinoma ex pleomorphic adenoma. It presents as a rapidly increasing parotid mass.

It has numerous histological appearances, such as sarcomatoid, invasive micropapillary, and oncocytic carcinomas. It is a very aggressive tumor with high incidence of local recurrence, regional and distant metastasis. It has poor prognosis, 55-56% died within 5 years [11].

Carcinoma Ex Pleomorphic Adenoma

Carcinoma ex pleomorphic adenoma is a wide-ranging group of carcinomas of the salivary glands [93]. It represents 12% of all salivary gland malignancies. They are most commonly arising from the parotid gland. About 7 to 27% develop after malignant degeneration of pleomorphic adenoma, typically after multiple times recurrent pleomorphic adenoma. However, the majority develop de novo. The types of the malignant component must be documented. The most commonly encountered one is carcinoma ex pleomorphic adenoma. The other categories are metastasizing pleomorphic adenoma and a true malignant mixed tumor, such as salivary duct carcinoma and myoepithelial carcinoma. It is more common in female patients with a peak incidence in the sixth and seventh decades of life [8].

The typical presentation is a preexisting parotid mass, which is rapidly growing and sometimes painful (Figure 7) [8, 11].

Carcinoma ex pleomorphic adenoma can be classified as [8, 11, 93]:

1. Intra-capsular tumor including carcinoma in situ
2. Minimally invasive when the tumor size is 1.5 to 6 mm
3. Widely-invasive when the tumor size is more than 1.5mm.

Local and distant metastasis might occur in 70% of the cases [11].

The outcome is good in intra-capsular and minimally invasive tumors. However, the widely invasive tumors have a high incidence of haematogenous spread. The cure rates at 5, 10, and 15 years are 40, 24, and 19 percent, respectively [8].

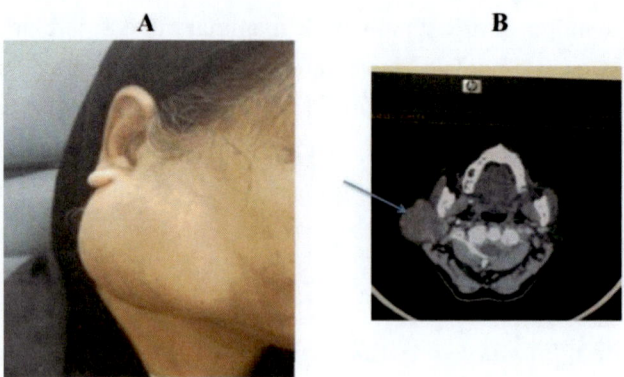

Figure 7 A and B. A 55-year-old lady presented with a long standing right parotid mass, which is, recently increases in size, worrisome for malignancy. An axial CT scan with contrast demonstrating a huge tumor involving the superficial lobe and extending into deep lobe (green arrow).

ROLE OF ELECTROPHYSIOLOGICAL MONITORING

Facial nerve injury during parotid surgery may result from nerve resection, injury to the vasa nervorum by thermal trauma and subsequent ischemia, compression, stretching, or ligature entrapment [94].

The majority of head and neck surgeons in the USA and United Kingdom were using facial nerve monitoring during parotid surgery [95-97]. It is an adjunctive tool that will help the surgeon in facial nerve main trunk identification when nerve localization is difficult despite the use of standard anatomic landmarks for nerve identification. This is could be due to altered nerve location by large, malignant, or deep lobe parotid tumor or due to fibrosis and adhesions in recurrent cases, prior radiation or chronic parotitis.

Moreover, it aids in identifying the peripheral nerve branch during retrograde facial nerve dissection, and to differentiate facial nerve from sensory nerve or non-nerve tissue during parotid dissection. In addition, it helps in the assessment of functional nerve integrity at the end of the surgery. This will assist in predicting the functional status of the nerve postoperatively and in patient counseling.

Normal response thresholds predict that the facial nerve function will be normal post operatively. Elevated response thresholds indicate variable degree of facial nerve paresis. However, the absence of nerve thresholds stimulation warns the surgeon about the possibility of either temporary or permanent facial nerve paralysis and will alarm them to look for any correctable cause [97-99].

In institutions where it's available, it's preferable to be used routinely in order to familiarize the surgeon with nerve monitoring system and to learn how to interpret the various signals. The operative time subsequently might be reduced with routine use of facial nerve monitoring. However, experienced surgeons may not believe in using it routinely because of their extremely low incidence of facial nerve paralysis [94].

Electrophysiological monitoring is relatively safe, with very rarely reported complications. Infection at the site of electrode placement and facial burn due to a technical deficit in a nerve monitoring device have been reported [100].

Some studies reported a statistically significant reduction in temporary and permanent facial paralysis and operative time reduction with facial nerve monitoring [101-103]. However, other studies did not show any correlation of intraoperative nerve responses with postoperative facial nerve function or any reduction in operative time [94, 101, 103].

Prospective, randomized controlled studies are required to answer the question as to whether facial nerve monitoring reduces the incidence of permanent facial paralysis or operative time during parotid surgery.

MANAGEMENT OF PAROTID NEOPLASMS

The mainstay of treatment is surgery. The extent of surgical resection varies from lumpectomy to radical parotidectomy, which is dependent on tumor location, (superficial versus deep lobe), size, and histology [5, 104].

The lesser the amount of parotid tissue resected, the lower the incidence of facial nerve dysfunction and Frey's syndrome. The reported rate of temporary facial nerve paresis is as high as 15% to 25% after superficial parotidectomy and 20% to 50% after total parotidectomy, whereas the overall rate of permanent facial nerve paresis is reported as 5% to 10% [45, 105-108].

Total parotidectomy is recommended in all high grade primary parotid tumors [5, 109]. For a superficial lobe low grade tumor with intact facial nerve, superficial parotidectomy or partial parotidectomy with an adequate margin of at least 1.5 cm is adequate [5, 110].

Superficial parotidectomy is considered the standard surgical procedure. It involves complete removal of the parotid gland superficial to the plane of the facial nerve. It is the treatment of choice for tumors in the superficial lobe, which are not involving the facial nerve.

Partial parotidectomy involves identification of the main trunk and only the branches surrounding the tumor. The parotid tumor is excised with a 0.5 to 1-cm cuff of normal surrounding parotid tissue [111].

Extracapsular dissection involves a dissection around the tumor capsule, without facial nerve identification and its performed under electromyographic monitoring of the facial nerve. The proponent of this surgical procedure stated that it offers better conditions for the facial nerve in revision cases as the nerve was not exposed during the primary surgery [112].

It has a low rate of facial nerve dysfunction and Frey's syndrome [113]. It is indicated in the cases of a superficially located mobile lesion or those arising from parotid tail and deep lobe parotid tumor, which are approached through transcervical approach [114].

Total conservative parotidectomy involves excision of entire parotid gland (superficial and deep lobes), while preserving the facial nerve. It is performed for: 1) tumors involving the deep lobe, with intact facial nerve functions; 2) high grade malignant tumors with a high risk for metastasis; 3) any parotid malignancy with an indication of metastasis to intraglandular or cervical lymph nodes; 4) any primary malignancy originating within the deep lobe itself; and 5) positive margin after superficial parotidectomy.

Total Parotidectomy with a sacrifice of facial nerve indications is similar to total conservative parotidectomy, when the nerve is involved by the tumor.

Radical parotidectomy implies excision of other structures than the parotid gland and facial nerve and is indicated when there is an involvement of the skin, infra-temporal fossa, mandible, and temporomandibular joint or petrous bone.

High grade malignant salivary gland tumors should undergo total parotidectomy [5, 109]. For a superficial lobe small low grade tumor with intact facial nerve, superficial parotidectomy or partial parotidectomy with an adequate margin of at least 1.5 cm is adequate [5, 110].

The facial nerve should be preserved if the facial nerve function is normal preoperatively. Facial nerve sacrifice does not offer better tumor control or survival benefit [5, 115, 116]. Sacrifice of the facial nerve or other structures is usually guided by surgical findings. If the nerve is non-functioning pre-operatively as result of tumor infiltiration, when cancer is found to surround the nerve or when there is direct invasion or change in the color and size of the nerve due to direct invasion, the nerve usually is excised [117, 118].

Facial nerve branches should be sacrificed only if the tumor is adherent to or surrounds the nerve, and if margins around the nerve are involved. In case of adenoid cystic carcinoma, if the sectioned nerve is involved, drilling of the temporal bone must be done until a free proximal stump is confirmed on frozen section. However, there is no certainty that the remaining nerve margin is free of tumor because of skip metastasis.

If there is a section of the facial nerve or one of its branches, it should be repaired immediately [119].

Modified Blair's incision is the standard incision for parotidectomy and is used for benign parotid tumors located in the superficial lobe. This incision was described by Blair in 1918 [120] and modified later on by Bailey in 1941 [121].

The incision is started just below the root of helix superiorly and extends in preauricular crease, then around the ear lobule over the mastoid tip. After that, it curves smoothly along the sternocleidomastoid muscle and then slightly forward in a natural skin crease in the upper neck (Figure 8).

It's a versatile incision and can be extended in the case an extensive surgery is required. The drawbacks are large incision and more likely to cause asymmetry as there is a greater area of dissection [120-122].

An alternative incision is a modified facelift incision in which the incision is placed behind the ear along the occipital hairline. The length of the incision posteriorly is dependent on the size of the tumor; the larger the tumor, the more likely the incision is extended posteriorly along the hairline. The exposure is very good and the neck scar camouflaged (Figure 9).

Both incisions can be extended or limited to a small incision, which is dependent on the extent of surgical resection.

The post surgical defect after resection of the tumor can be reconstructed and various techniques have been described such as rotation of the upper part of sternocleidomastoid muscle, approximation of superficial aponeurtic fascia, and placement of fat or alloderm.

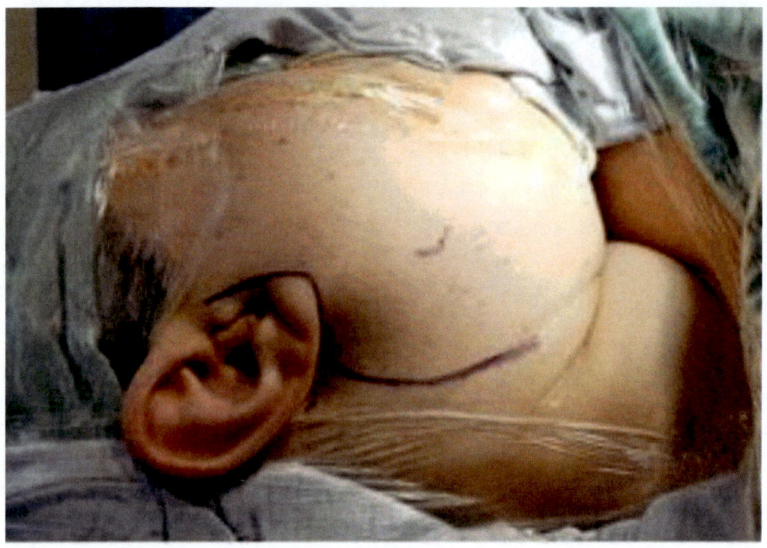

Figure 8. Modified Blair incision.

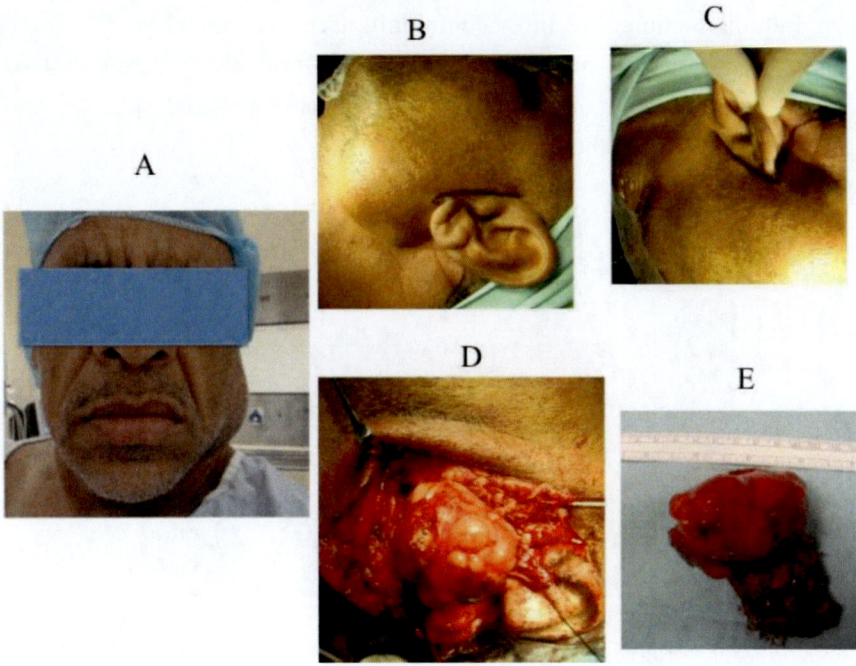

Figure 9. A 58-year-old gentleman presented with a large left parotid mass as shown in image A. He underwent tumor excision using a modified face lift incision, that was extended along the hairline posteriorly as shown in image B &C. Image D demonstrating the good exposure provided by the incision. Image E showed the excised tumor that was around 4.5 x 3 cm in size.

For the excision of deep lobe parotid tumors, several approaches can be used, such as transparotid approach, transparotid approach with mandibulotomy, transcervical approach, or a combined transparotid transcervical approach which is dependent on tumor extent [123]. For the transparotid approach, modified Blair or face-lift incision can be used.

Facial Nerve Identification during Parotid Surgery

The parotid gland is divided arbitrarily into superficial and deep lobe by the facial nerve [124]. Parotid surgery is an anatomical dissection. The facial nerve is identified by two approaches: the conventional antegrade or by retrograde dissection. The most commonly used anatomical landmarks to identify facial nerve trunk are stylomastoid foramen, tympanomastoid suture, posterior belly of digastric, tragal pointer, mastoid process, and peripheral branches of the facial nerve.

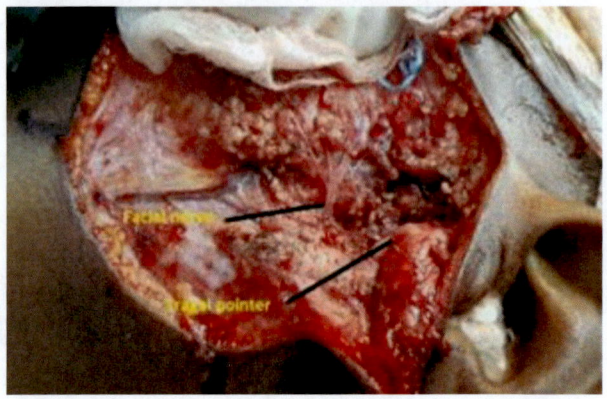

Figure 10. Demonstrating the location of the facial nerve main trunk in relation to the tragal pointer, which is inferior and medial.

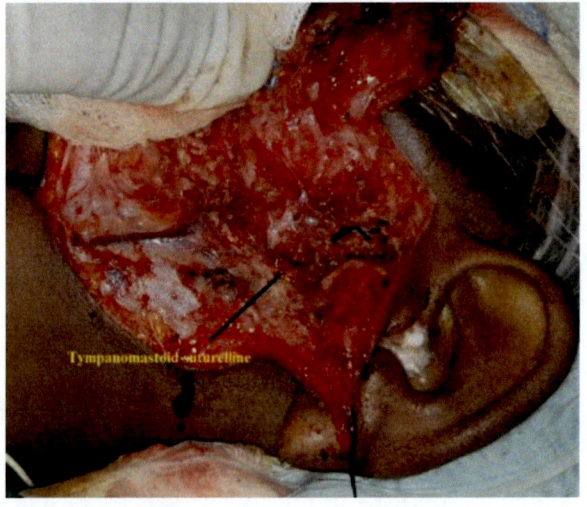

Figure 11. Demonstrating the location of the facial nerve main trunk in relation to the tympanomastoid suture (most constant landmark), which is deep to the suture.

The nerve is located approximately 16.36 - 16.61 mm deep and inferior to tragal pointer (Figure10). The most constant landmark for facial nerve identification is the tympanomastoid suture line (Figure 11). The nerve is located deep to it at an average distance of 3.5 - 3.87 mm in a cadaver and in live patient dissection [125].

The stylomastoid foramen is located medial to the insertion of posterior belly of digastric muscle on the mastoid tip, and the nerve is found to be at a mean distance of 7.5- 8.03 mm medially [125].

In a retrograde dissection for main trunk identification, one or more of the peripheral branches can be identified distally and followed medially to the main trunk. The marginal mandibular nerve and buccal branches are the most commonly used branches for facial nerve identification.

ROLE OF INTRAOPERATIVE FROZEN SECTION

Frozen section on the operative specimen could be done for these purposes: 1) to clarify pre-operative diagnosis; 2) to check surgical margins; 3) to determine whether facial nerve or neck node involvement is present; and 4) to check margins and assess nerve involvement with a pre-/surgical malignant diagnosis.

EUROPEAN SALIVARY GLAND SOCIETY PAROTID SURGERY CLASSIFICATION

The European Salivary Gland Society (ESGS) has published a classification using multiple levels of dissection similar to neck dissection classification in order to classify the parotid resection procedures (Table 3) [104].

Table 3. Classification of Parotid Resection types by the European Salivary Gland Society [104]

Resection	Resected levels and other factors
Superficial superior lobe	I
Superficial inferior lobe	II
Deep inferior lobe	III
Deep superior lobe	IV
Accessory parotid types	V
Parotidectomy	Dissection of the facial nerve + resection ≥1 level
Formal Parotidectomy types	
Lateral Parotidectomy	Level I + II
Total Parotidectomy	Level I - IV
Partial Parotidectomy types	
Partial lateral Parotidectomy	Level I or II
Selective deep lobe resection	Level III or IV
Extracapsular dissection	No dissection of the facial nerve and/or resection <1 level

In this classification, the parotid gland is divided into five levels. The superior level is the area corresponding to the temporofacial division of the facial nerve, and the inferior level corresponds to the cervicofacial division. Level I corresponds to parotid tissue found lateral to the plane of the facial nerve and cranial to an imaginary line drawn from the facial nerve trunk to Stensen's duct.

Level II represents parotid tissue in the superficial lobe that is caudal to the above mentioned imaginary line. For the deep lobe, glandular element inferior to the imaginary line is labeled as Level III and, superiorly, Level IV. The accessory lobe is regarded as Level V (Figure 12) [104]. Based on this classification, partial parotidectomy when a surgeon removed only part of level I and part of level II and according to ESGS classification will be expressed as level I and II, in which the entire superficial lobe is resected. Therefore, it is not accurately described by the ESGS classification system. Subsequently, a modified classification was proposed by Wong and colleagues (2017) [126] in which parotid Level I was subdivided into two additional sublevels: IA and IB. The superior (temporofacial) division of facial nerve was used as the anatomic landmark to subdivide Level I into Sublevels IA and IB.

In addition, parotid Level II was subdivided into sublevels IIA and IIB using the inferior (cervicofacial) division of facial nerve as a landmark (Figure 13).

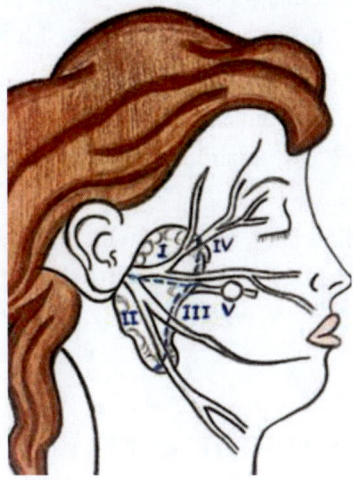

Figure 12. ESGS classification demonstrating parotid gland five levels classification.

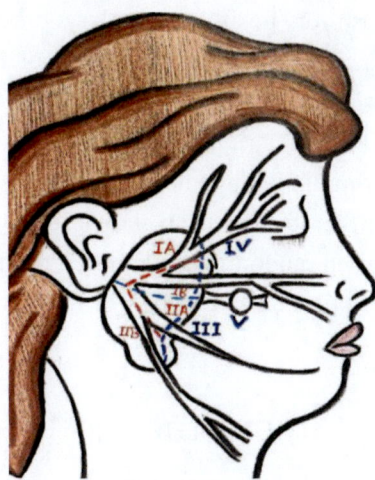

Figure 13. ESGS classification demonstrating parotid gland sub-levels classification.

SURGICAL MANAGEMENT OF THE NECK METASTASES

The reported incidence of positive lymph nodes metastasis has ranged from 13-39%. [127]. Neck dissection should be performed if there is clinical or radiological evidence of nodal disease, high grade tumor (i.e., adenocarcinoma, squamous and undifferentiated carcinomas, high-grade mucoepidermoid carcinoma, and carcinoma ex pleomorphic adenoma) [128, 129], advanced T-stage, extraglandular extension of the tumor, or facial palsy [5, 130, 131].

Therapeutic neck dissection should include dissection of level I-V (5). In an elective neck dissection at least level I-III should be included [108, 132].

RADIOTHERAPY (RT)

Adjuvant RT is recommended in recurrent multifocal pleomorphic adenoma, malignant large tumors (>4 cm), high grades tumors, adenoid cystic carcinoma, close or positive margins, presence of peri-neural invasion, extraglandular extension, and presence of nodal and metastatic disease [108, 133].

Definitive radical RT is restricted to unresectable tumors. This type of treatment is usually for palliative intent.

Fast neutron beam therapy has been shown to be more beneficial than standard photon therapy in randomized controlled trials. However, its use is limited by the extremely limited availability of fast neutron RT units [134].

CHEMOTHERAPY

Chemotherapy is not indicated in standard curative treatment of primary parotid cancer and it does not influence the survival compared to radiotherapy alone [5, 135].

It has a role only in palliative setting in patients with recurrent unresectable disease or distant metastases. It may have a palliative benefit for a small proportion of patients with recurrent/metastatic adenoid cystic carcinomas after due consideration of other modalities (palliative radiation, metastatectomy of solitary lesions).

TARGETED THERAPY

Salivary gland tumors may have molecular targets like c-kit, EGFR, and Her-2. However, targeted therapies do not show any definite antitumor activity and clinical trials are restricted to phase II trials until now [5, 136, 137].

COMPLICATIONS

The complications expected after benign parotid tumor surgery are few and the patients are expected to have normal function postoperatively. The postoperative complications can be categorized as intraoperative and postoperative complications.

The intraoperative complications include resection of one or more branches of facial nerve, rupture of tumor capsule, and incomplete tumor resection.

The resection of facial nerve branches, if developed, should be managed at the same time by doing end-to-end anastomosis by suturing the epineurium using 8/0 vicryl if there is no tension or by putting fibrin glue. If there is a tension between the two resected ends, then a cable nerve graft from great auricular nerve should be used [138-142].

Rupture of tumor capsule should be avoided, and if it happens the surgical bed should be washed thoroughly with any type of irrigating solution however there is no study to prove the efficacy of this. Opponents for this debate that washing might spread the viable cells in the surgical bed [5, 143].

Rupture of the tumor during surgery might lead to a higher than 14-fold increase in the risk of recurrence [144]. However, others showed no evidence of recurrence [145].

The postoperative complications are as the following:

Facial Nerve Dysfunction

Facial nerve dysfunction is one of the most common complications encountered. Postoperative temporary and permanent facial nerve paralysis reported to be 9.1% to 64.0% and 0% to 3.9% [146-148].

The incidence is directly related to the extent of surgical resection due to the stretching of the nerve or trauma to the vasa nervorum. It is more frequent with revision surgeries and congenital lesions. Temporary facial nerve dysfunction usually recovers within two weeks or up to 6 months. Eye protection is essential in those with incomplete eye closure by using ophthalmic solution and eye protection at night. In addition, a temporary ptosis should be performed, by doing partial tarsorrhaphy or botulinum toxin injection [138, 140, 141, 145, 149-151].

Ear Lobule Hypoesthesia

Ear lobule hypoesthesia is due to section of the posterior branches of the greater auricular nerve and the area of numbness mostly will recover within one year of the surgery in majority of the cases. However, small areas of numbness may persist. This is an avoidable complication in most of the cases by preserving the posterior branches of the greater auricular nerve whenever possible [152].

Greater Auricular Nerve Neuroma

Greater auricular nerve neuroma occurred due to injury to the greater auricular nerve and can be managed by excising it.

Cosmetic Complications

Surgical site depression is dependent on the amount of the resected parotid tissue. It can be reduced by suturing the superficial musculoaponeurotic system (SMAS) to the preauricular tissue and placement of sterenocledomastoid flap or alloderm [153, 154].

Skin flap necrosis, especially the postauricular part of the incision, can be avoided by a proper design of the flap.

A Keloid or hypertrophied scar might develop. This is can be managed by steroid injection or by revising the scar and injecting steroid [138].

Haematoma develops due to inadequate haemostasis. It should be managed by draining it and controlling the bleeding site.

Sialocele can be treated by aspiration and applying pressure dressing.

Salivary fistula should be treated by putting pressure dressing. In case there is no improvement, botulinum toxin injection should be used.

Frey syndrome or gustatory sweating:

Frey syndrome is characterized by unilateral sweating and flushing of facial skin in the parotid gland area that occur while eating. It has been reported as high as 50% [138].

The pathogenesis of Frey syndromeis is based on the aberrant regeneration of sectioned parasympathetic secretomotor fibres of the auriculotemporal nerve with inappropriate innervation of the cutaneous facial sweat glands that are normally innervated by sympathetic cholinergic fibres It can be diagnosed by a minor starch test. This complication can be avoided by raising a thick skin flap during parotid surgery or by placing a muscle or SMAS flap. The non surgical treatment of this complication includes local application of anticholinergic agents, such as scopolamine or glycopyrrolate. The surgical options include tympanic neuroectomy, placement of a barrier between the skin flap and parotid bed such as sternocleidomastoid muscle flap, and facia lata or alloderm. Recently, local injection of botulinum toxin in the affected area showed good results. It acts by preventing the release of acetylcholine at the neuromuscular junction and resulting in chemical denervation. That subsequently leads to the paralysis of striated muscle and eccrine glands [146, 155-159].

RECURRENT PLEOMORPHIC ADENOMA

Recurrent pleomorphic adenoma is one of the challenging head and neck surgeries to perform as its mostly multifocal (33 to 98%) [146, 160].

It appears in relatively young patients and up to 15 years after the initial surgery. It's associated with an increase in postoperative facial nerve paralysis, further recurrences, and malignant degeneration [146, 160].

The pleomorphic adenoma recurrence causes were divided by Dulguerov et al. (2017) into pathology and surgery related causes as shown in Table 1 [161, 162].

The size of these recurred nodules were less than 1mm in most cases. This would explain the cause of multiple recurrences. Multiple recurrences were reported in 52% of the cases with a 20 years median follow up after the primary surgery [163].

Magnetic resonance imaging is the best imaging in determining the extent of recurrent pleomorphic adenoma. It appears as rounded and of low intensity on T1-weighted MRI and of high Intensity on T2-weighted MRI and there is a mild enhancement with contrast administration [138].

Extent of surgical resection is important in determining the recurrence rate. The recurrence rate has been reported as high as 45% with enculation [164-166].

It is reduced with superficial parotidectomy to 1%–4% and to 0%–0.4% with total parotidectomy [151].

Total parotidectomy followed by radiation therapy is the best treatment for reducing the recurrence and providing better local control. However, surgical extent can be modified according to recurrence extent (Figure 14).

Postoperative temporary and permanent facial nerve paralysis is reported to be 9.1% to 64.0% and 0% to 3.9% [148-150]. However, the incidence increases with revision surgery to 90%-100% for temporary paralysis and 11.3%-40.0% for permanent paralysis [146, 163, 167-169].

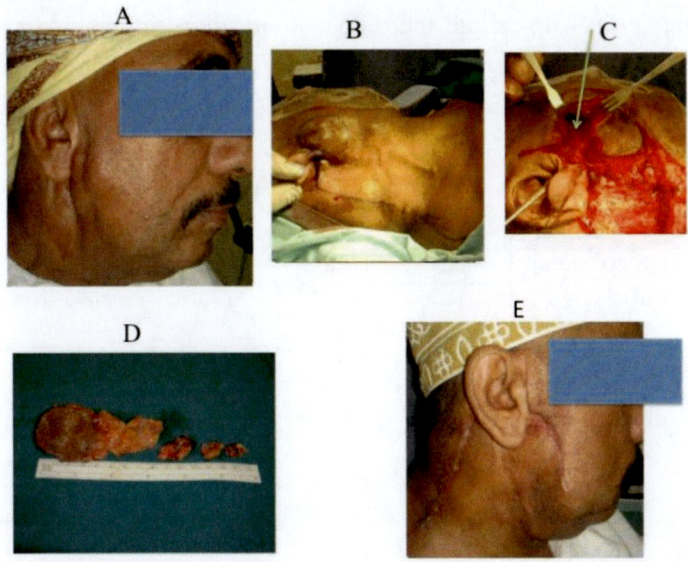

Figure 14. A - A 55 year- old gentleman presented with a recurrent pleomorphic adenoma which was adherent to skin. B - Skin marking of the tumor that was adherent to skin and the outline of cervical rotation flap that used to reconstruct the defect. C - Temporofacial division of facial nerve is intact (green arrow) but the cervicofacial division is surrounded by the recurrent tumor and sacrificed. D - Excised tumor with multiple small recurrent nodules. E - Postoperative appearance 2 months after the surgery.

Using nerve monitoring in recurrent pleomorphic adenoma surgery may help in reducing the morbidity of facial nerve injury [161, 170, 171].

It reduces operative time and incidence of facial nerve paralysis. In addition, it reduces the recovery time of temporary facial nerve dysfunction as the surgeon tends to preserve the fibrosis that surround the nerve rather than skeletonize it in order to reduce the mechanical trauma.

The reported incidence of permanent facial nerve paralysis after parotidectomy in unmonitored patients was 23.3% compared to 10.7% in monitored cases [172].

In recurrent pleomorphic adenoma parotid surgery, the incidence of temporary and permanaent facial nerve dysfunction are higher compared to primary surgery due to presence of fibrosis and mechanical surgical trauma to the nerve as dissecting the nerve from the fibrosis may lead to devasularization and perineurium exposure. The use of nerve monitoring will lead to nerve identification without exposing it [148].

Table 4. Causes of recurrence in pleomorphic adenoma

Causes of recurrence in pleomorphic adenoma	Pathology- related	Surgery- related
	Thin capsule or lack of capsule	Tumor rupture
	Pseudopodia	Insufficient resection margins
	Satellite nodules	Incomplete tumor resection
	Multi-centricity	

The local control rate after the first redo surgery ranged from 36 to 98% and 43% to 45% after second redo surgery [146, 160, 163, 167, 172].

Adjuvant radiation is indicated in cases of multifocal recurrences, after multiple recurrences, when complete resection of recurrent lesion is not possible and when there is a need to sacrifice the facial nerve [173, 174].

The reported local control rate of combination of surgery and radiation ranged from 56-100% compared to 5.8-71% in those who were treated with surgery alone [175-180].

The malignant transformation of recurrent pleomorphic adenoma to carcinoma ex-pleomorphic adenoma has been reported in the range of 0 to 23% [146, 163, 167, 181, 182].

REFERENCES

[1] Zhan, Kevin Y., Sobia F. Khaja, Allen B. Flack, and Terry A. Day. "Benign Parotid Tumors." *Otolaryngologic Clinics of North America* 49, no. 2 (2016): 327–42. https://doi.org/10.1016/j.otc.2015.10.005.

[2] Brennan, Pa, M Ammar, and J Matharu. "Contemporary Management of Benign Parotid Tumours - the Increasing Evidence for Extracapsular Dissection." *Oral Diseases* 23, no. 1 (November 2016): 18–21. https://doi.org/10.1111/odi.12518.

[3] Kadletz, Lorenz, Stefan Grasl, Matthäus C. Grasl, Christos Perisanidis, and Boban M. Erovic. "Extracapsular Dissection versus Superficial Parotidectomy in Benign Parotid Gland Tumors: The Vienna Medical School Experience." *Head & Neck* 39, no. 2 (May 2016): 356–60. https://doi.org/10.1002/hed.24598.

[4] Patel, Depak K., and Randall P. Morton. "Demographics of Benign Parotid Tumours: Warthin's Tumour versus Other Benign Salivary Tumours." *Acta Oto-Laryngologica* 136, no. 1 (August 2015): 83–86. https://doi.org/10.3109/00016489.2015.1081276.

[5] Thielker, Jovanna, Maria Grosheva, Stephan Ihrler, Andrea Wittig, and Orlando Guntinas-Lichius. "Contemporary Management of Benign and Malignant Parotid Tumors." *Frontiers in Surgery* 5 (November 2018). https://doi.org/10.3389/fsurg.2018.00039.

[6] Bjørndal, Kristine, Annelise Krogdahl, Marianne Hamilton Therkildsen, Jens Overgaard, Jørgen Johansen, Claus A. Kristensen, Preben Homøe, et al. "Salivary Gland Carcinoma in Denmark 1990–2005: A National Study of Incidence, Site and Histology. Results of the Danish Head and Neck Cancer Group (DAHANCA)." *Oral Oncology* 47, no. 7 (2011): 677–82. https://doi.org/10.1016/j.oraloncology.2011.04.020.

[7] Xiao, Christopher C., Kevin Y. Zhan, Shai J. White-Gilbertson, and Terry A. Day. "Predictors of Nodal Metastasis in Parotid

Malignancies." *Otolaryngology–Head and Neck Surgery* 154, no. 1 (2015): 121–30. https://doi.org/10.1177/0194599815607449.

[8] Sood, S, M Mcgurk, and F Vaz. "Management of Salivary Gland Tumours: United Kingdom National Multidisciplinary Guidelines." *The Journal of Laryngology & Otology* 130, no. S2 (2016). https://doi.org/10.1017/s0022215116000566.

[9] Chan, John K. C., Adel K. El-Naggar, Jennifer Rubin Grandis, Pieter Johannes Slootweg, and Takashi Takata. *WHO Classification of Head and Neck Tumours*. Lyon: International Agency for Research on Cancer, 2017.

[10] Ihrler S, Agaimy A, Guntinas-Lichius O, Mollenhauer M. WHO-Klassifikation 2017: Neues zu Tumoren der Speicheldrüsen. *Pathologe* (2018).

[11] Slootweg, P. J., and A. K. El-Naggar. "World Health Organization 4th Edition of Head and Neck Tumor Classification: Insight into the Consequential Modifications." *Virchows Archiv* 472, no. 3 (2018): 311–13. https://doi.org/10.1007/s00428-018-2320-6.

[12] Seethala, Raja R., and Göran Stenman. "Update from the 4th Edition of the World Health Organization Classification of Head and Neck Tumours: Tumors of the Salivary Gland." *Head and Neck Pathology* 11, no. 1 (2017): 55–67. https://doi.org/10.1007/s12105-017-0795-0.

[13] Brennan, P. A., M. K. Herd, D. C. Howlett, D. Gibson, and R. S. Oeppen. "Is Ultrasound Alone Sufficient for Imaging Superficial Lobe Benign Parotid Tumours before Surgery?" *British Journal of Oral and Maxillofacial Surgery* 50, no. 4 (2012): 333–37. https://doi.org/10.1016/j.bjoms.2011.01.018.

[14] Lee, Y. Y. P., K. T. Wong, A. D. King, and A. T. Ahuja. "Imaging of Salivary Gland Tumours." *European Journal of Radiology* 66, no. 3 (2008): 419–36. https://doi.org/10.1016/j.ejrad.2008.01.027.

[15] Fakhry, N., F. Antonini, J. Michel, M. Penicaud, J. Mancini, A. Lagier, L. Santini, et al. "Fine-Needle Aspiration Cytology in the Management of Parotid Masses: Evaluation of 249 Patients." *European Annals of Otorhinolaryngology, Head and Neck Diseases*

129, no. 3 (2012): 131–35. https://doi.org/10.1016/j.anorl.2011.10.008.

[16] Zbären, Peter, Dominique Guélat, Heinz Loosli, and Edouard Stauffer. "Parotid Tumors: Fine-Needle Aspiration and/or Frozen Section." *Otolaryngology–Head and Neck Surgery* 139, no. 6 (2008): 811–15. https://doi.org/10.1016/j.otohns.2008.09.013.

[17] Atula, Timo, Reidar Grénman, Pekka Laippala, and Pekka-Juhani Klemi. "Fine-Needle Aspiration Biopsy in the Diagnosis of Parotid Gland Lesions: Evaluation of 438 Biopsies." *Diagnostic Cytopathology* 15, no. 3 (1996): 185–90. https://doi.org/10.1002/(sici)1097-0339(199609)15:3<185::aid-dc2>3.0.co;2-g.

[18] Stewart, C. J. R., K. Mackenzie, G. W. Mcgarry, and A. Mowat. "Fine-Needle Aspiration Cytology of Salivary Gland: A Review of 341 Cases." *Diagnostic Cytopathology* 22, no. 3 (January 2000): 139–46. https://doi.org/10.1002/(sici)1097-0339(20000301)22:3<139::aid-dc2>3.0.co;2-a.

[19] Postema, Rolf J., Mari-Louise F. Van Velthuysen, Michiel W. M. Van Den Brekel, Alfons J. M. Balm, and Johannes L. Peterse. "Accuracy of Fine-Needle Aspiration Cytology of Salivary Gland Lesions in the Netherlands Cancer Institute." *Head & Neck* 26, no. 5 (2004): 418–24. https://doi.org/10.1002/hed.10393.

[20] Cohen, Erik G., Snehal G. Patel, Oscar Lin, Jay O. Boyle, Dennis H. Kraus, Bhuvanesh Singh, Richard J. Wong, Jatin P. Shah, and Ashok R. Shaha. "Fine-Needle Aspiration Biopsy of Salivary Gland Lesions in a Selected Patient Population." *Archives of Otolaryngology–Head & Neck Surgery* 130, no. 6 (January 2004): 773. https://doi.org/10.1001/archotol.130.6.773.

[21] Bajaj, Y, S Singh, N Cozens, and J Sharp. "Critical Clinical Appraisal of the Role of Ultrasound Guided Fine Needle Aspiration Cytology in the Management of Parotid Tumours." *The Journal of Laryngology & Otology* 119, no. 4 (2005): 289–92. https://doi.org/10.1258/0022215054020421.

[22] Seethala, Raja R., Virginia A. Livolsi, and Zubair W. Baloch. "Relative Accuracy of Fine-Needle Aspiration and Frozen Section in

the Diagnosis of Lesions of the Parotid Gland." *Head & Neck* 27, no. 3 (2005): 217–23. https://doi.org/10.1002/hed.20142.

[23] Aversa, Salvatore, Cristina Ondolo, Enrico Bollito, Gianluca Fadda, and Salvatore Conticello. "Preoperative Cytology in the Management of Parotid Neoplasms." *American Journal of Otolaryngology* 27, no. 2 (2006): 96–100. https://doi.org/10.1016/j.amjoto.2005.07.015.

[24] Uguz M. Z., Onal H. K., Eroglu O. O., et al. Sensitivity and specificity of fine-needle aspiration biopsy in parotid masses. *Kulak Burun Bogaz Ihtis Derg* 2007;17:96—9.

[25] Hernández, Álvaro Antonio Herrera, Julio Alexander Díaz Pérez, Carlos Andrés García, Loren Paola Herrera, Paola Aranda Valderrama, and Luis Carlos Orozco Vargas. "Evaluation of Fine Needle Aspiration Cytology in the Diagnosis of Cancer of the Parotid Gland." *Acta Otorrinolaringologica (English Edition)* 59, no. 5 (2008): 212–16. https://doi.org/10.1016/s2173-5735(08)70225-4.

[26] Lim, Chwee Ming, Juliana They, Kwok Seng Loh, Siew Shuen Chao, Lynne Hseuh Yee Lim, and Luke Kim Siang Tan. "Role Of Fine-Needle Aspiration Cytology In The Evaluation Of Parotid Tumours." *ANZ Journal of Surgery* 77, no. 9 (2007): 742–44. https://doi.org/10.1111/j.1445-2197.2007.04222.x.

[27] Lin, Aaron C., and Neil Bhattacharyya. "The Utility of Fine Needle Aspiration in Parotid Malignancy." *Otolaryngology–Head and Neck Surgery* 136, no. 5 (2007): 793–98. https://doi.org/10.1016/j.otohns.2006.12.026.

[28] Carrillo, José F., Rene Ramírez, Lorena Flores, Margarita C. Ramirez-Ortega, Myrna D. Arrecillas, Margarita Ibarra, Rita Sotelo, Sergio Ponce-De-León, and Luis F. Oñate-Ocaña. "Diagnostic Accuracy of Fine Needle Aspiration Biopsy in Preoperative Diagnosis of Patients with Parotid Gland Masses." *Journal of Surgical Oncology* 100, no. 2 (January 2009): 133–38. https://doi.org/10.1002/jso.21317.

[29] Jafari, Alice, Benedicte Royer, Marine Lefevre, Pascal Corlieu, Sophie Périé, and Jean Lacau St Guily. "Value of the Cytological Diagnosis in the Treatment of Parotid Tumors." *Otolaryngology–*

Head and Neck Surgery 140, no. 3 (2009): 381–85. https://doi.org/10.1016/j.otohns.2008.10.032.

[30] Schmidt, Robert L., Brian J. Hall, Andrew R. Wilson, and Lester J. Layfield. "A Systematic Review and Meta-Analysis of the Diagnostic Accuracy of Fine-Needle Aspiration Cytology for Parotid Gland Lesions." *American Journal of Clinical Pathology* 136, no. 1 (2011): 45–59. https://doi.org/10.1309/ajcpoie0cznat6sq.

[31] Fakhry, N., F. Antonini, J. Michel, M. Penicaud, J. Mancini, A. Lagier, L. Santini, et al. "Fine-Needle Aspiration Cytology in the Management of Parotid Masses: Evaluation of 249 Patients." *European Annals of Otorhinolaryngology, Head and Neck Diseases* 129, no. 3 (2012): 131–35. https://doi.org/10.1016/j.anorl.2011.10.008.

[32] Obrien, Christopher J. "Current Management of Benign Parotid Tumors? The Role of Limited Superficial Parotidectomy." *Head & Neck* 25, no. 11 (2003): 946–52. https://doi.org/10.1002/hed.10312.

[33] Romano, Erica B., Jason M. Wagner, Anthony M. Alleman, Lichao Zhao, Rachel D. Conrad, and Greg A. Krempl. "Fine-Needle Aspiration with Selective Use of Core Needle Biopsy of Major Salivary Gland Tumors." *The Laryngoscope* 127, no. 11 (2017): 2522–27. https://doi.org/10.1002/lary.26643.

[34] Witt, Benjamin L., and Robert L. Schmidt. "Ultrasound-Guided Core Needle Biopsy of Salivary Gland Lesions: A Systematic Review and Meta-Analysis." *The Laryngoscope* 124, no. 3 (2013): 695–700. https://doi.org/10.1002/lary.24339.

[35] Liu, C. Carrie, Ashok R. Jethwa, Samir S. Khariwala, Jonas Johnson, and Jennifer J. Shin. "Sensitivity, Specificity, and Posttest Probability of Parotid Fine-Needle Aspiration." *Otolaryngology–Head and Neck Surgery* 154, no. 1 (2015): 9–23. https://doi.org/10.1177/0194599815607841.

[36] Amin MB, Edge S, Greene F, Byrd DR, Brookland RK, Washington MK, Gershenwald JE, Compton CC, Hess KR, et al. (Eds.). *AJCC Cancer Staging Manual (8th edition).* Springer International

Publishing: American Joint Commission on Cancer; 2017 (cited 2016 Dec 28).

[37] Eveson, J. W., and R. A. Cawson. "Salivary Gland Tumours. A Review of 2410 Cases with Particular Reference to Histological Types, Site, Age and Sex Distribution." *The Journal of Pathology* 146, no. 1 (1985): 51–58. https://doi.org/10.1002/path.1711460106.

[38] Spiro, Ronald H. "Salivary Neoplasms: Overview of a 35-Year Experience with 2,807 Patients." *Head & Neck Surgery* 8, no. 3 (1986): 177–84. https://doi.org/10.1002/hed.2890080309.

[39] Guerra, Germano, Domenico Testa, Stefania Montagnani, Domenico Tafuri, Francesco Antonio Salzano, Aldo Rocca, Bruno Amato, et al. "Surgical Management of Pleomorphic Adenoma of Parotid Gland in Elderly Patients: Role of Morphological Features." *International Journal of Surgery* 12 (2014). https://doi.org/10.1016/j.ijsu.2014.08.391.

[40] Stennert, Eberhard, Orlando Guntinas-Lichius, Jens Peter Klussmann, and Georg Arnold. "Histopathology of Pleomorphic Adenoma in the Parotid Gland: A Prospective Unselected Series of 100 Cases." *The Laryngoscope* 111, no. 12 (2001): 2195–2200. https://doi.org/10.1097/00005537-200112000-00024.

[41] Zbären, Peter, and Edouard Stauffer. "Pleomorphic Adenoma of the Parotid Gland: Histopathologic Analysis of the Capsular Characteristics of 218 Tumors." *Head & Neck* 29, no. 8 (2007): 751–57. https://doi.org/10.1002/hed.20569.

[42] Takahashi, Mitsuaki, Kazuhiko Hokunan, and Tokuji Unno. "Immunohistochemical Study of Basement Membrane in Pleomorphic Adenomas of the Parotid Gland." *Nippon Jibiinkoka Gakkai Kaiho* 95, no. 11 (1992). https://doi.org/10.3950/jibiinkoka.95.1759.

[43] Takahashi, Mitsuaki, Megumi Kumai, Toshihiko Kamito, Motoharu Uehara, and Tokuji Unno. "Clinico-Pathological Findings of Recurrent Pleomorphic Adenomas of the Parotid Gland." *Nippon Jibiinkoka Gakkai Kaiho* 94, no. 4 (1991): 489–94. https://doi.org/10.3950/jibiinkoka.94.489.

[44] Lawson, H. H. "Capsular Penetration and Perforation in Pleomorphic Adenoma of the Parotid Salivary Gland." *British Journal of Surgery* 76, no. 6 (1989): 594–96. https://doi.org/10.1002/bjs.1800760622.

[45] Mcgurk, M., A. Renehan, E. N. Gleave, and B. D. Hancock. "Clinical Significance of the Tumour Capsule in the Treatment of Parotid Pleomorphic Adenomas." *British Journal of Surgery* 83, no. 12 (1996): 1747–49. https://doi.org/10.1002/bjs.1800831227.

[46] Hancock, B. D. "Pleomorphic Adenomas of the Parotid: Removal without Rupture." *Annals of the Royal College of Surgeons of England* 69 (1987): 293–95. https://www.ncbi.nlm.nih.gov/pmc/articles/PMC2498527/.

[47] Orita, Yorihisa, Kazuo Hamaya, Kentaroh Miki, Akiko Sugaya, Misato Hirai, Kiyoko Nakai, Sohichiroh Nose, and Tadashi Yoshino. "Satellite Tumors Surrounding Primary Pleomorphic Adenomas of the Parotid Gland." *European Archives of Oto-Rhino-Laryngology* 267, no. 5 (July 2009): 801–6. https://doi.org/10.1007/s00405-009-1149-7.

[48] Yoo, George H., David W. Eisele, Frederic B. Askin, Jeffrey S. Driben, and Michael E. Johns. "Warthin??s Tumor." *The Laryngoscope* 104, no. 7 (1994). https://doi.org/10.1288/00005537-199407000-00004.

[49] Chulam, T. C., A. L. Noronha Francisco, J. Goncalves Filho, C. A. Pinto Alves, L. P. Kowalski. Warthin's tumour of the parotid gland:our experience. *ACTA otorhinolaryngologica ita lica* 2013; 33:393-397.

[50] de Ru J. A., Plantinga R. F., Majoor M. H., van Benthem P. P., Slootweg P. J., Peeters P. H., and Hordijk G. J. "Warthin's Tumour and Smoking." *B-ENT* 1(2):63-6 (2005). https://www.ncbi.nlm.nih.gov/pubmed/16044736.

[51] Faur, Alexandra, Elena Lazăr, Mărioara Cornianu, Alis Dema, Camelia Gurban Vidita, and Atena Găluşcan. "Warthin Tumor: a Curious Entity – Case Reports and Review of Literature." *Romanian Journal of Morphology and Embryology* 50, no. 2 (2009): 296–73.

https://www.researchgate.net/publication/24419789_Warthin_tumor_A_curious_entity_-_Case_reports_and_review_of_literature.

[52] Maiorano, E., L. Lo Muzio, G. Favia, and A. Piattelli. "Warthins Tumour: a Study of 78 Cases with Emphasis on Bilaterality, Multifocality and Association with Other Malignancies." *Oral Oncology* 38, no. 1 (2002): 35–40. https://doi.org/10.1016/s1368-8375(01)00019-7.

[53] Guntinas-Lichius, Orlando, Bettina Gabriel, and J. Peter Klussmann. "Risk of Facial Palsy and Severe Freys Syndrome after Conservative Parotidectomy for Benign Disease: Analysis of 610 Operations." *Acta Oto-Laryngologica* 126, no. 10 (2006): 1104–9. https://doi.org/10.1080/00016480600672618.

[54] Teymoortash, A., Y. Krasnewicz, and J. A. Werner. "Clinical Features of Cystadenolymphoma (Warthin's Tumor) of the Parotid Gland: A Retrospective Comparative Study of 96 Cases." *Oral Oncology* 42, no. 6 (2006): 569–73. https://doi.org/10.1016/j.oraloncology.2005.10.017.

[55] Mcgarry, James G., Maeve Redmond, John B. Tuffy, Lorraine Wilson, and Seamus Looby. "Metastatic Pleomorphic Adenoma to the Supraspinatus Muscle: a Case Report and Review of a Rare Aggressive Clinical Entity." *Journal of Radiology Case Reports* 9, no. 10 (2015). https://doi.org/10.3941/jrcr.v9i10.2283.

[56] Bradley, Patrick J. "???Metastasizing Pleomorphic Salivary Adenoma??? Should Now Be Considered a Low-Grade Malignancy with a Lethal Potential." *Current Opinion in Otolaryngology & Head and Neck Surgery* 13, no. 2 (2005): 123–26. https://doi.org/10.1097/01.moo.0000153450.87288.2a.

[57] Knight, James, and Kumaran Ratnasingham. "Metastasising Pleomorphic Adenoma: Systematic Review." *International Journal of Surgery* 19 (2015): 137–45. https://doi.org/10.1016/j.ijsu.2015.04.084.

[58] Bonet-Loscertales, M., M. Armengot-Carceller, J. Gaona-Morales, and J. Basterra-Alegria. "Multicentric Recurrent Parotid Pleomorphic

Adenoma in a Child." *Medicina Oral Patología Oral y Cirugia Bucal*, 2010. https://doi.org/10.4317/medoral.15.e743.

[59] Soteldo, Javier, and Nathasha Aranaga. "Metastasizing Pleomorphic Adenoma of the Parotid Gland." *Ecancermedicalscience* 11 (2017). https://doi.org/10.3332/ecancer.2017.758.

[60] Thompson, Lester D. R. "World Health Organization Classification of Tumours: Pathology and Genetics of Head and Neck Tumours." *Ear, Nose & Throat Journal* 85, no. 2 (2006): 74–74. https://doi.org/10.1177/014556130608500201.

[61] Rodríguez-Fernández J., Mateos-Micas M., Martínez-Tello F., Berjón J., Montalvo J., Forteza-González G., and Galan-Hernández R. "Metastatic Benign Pleomorphic Adenoma. Report of a Case and Review of the Literature." *Med Oral Patol Oral Cir Bucal* 13(3):E193-6 (March 1, 2008). https://www.ncbi.nlm.nih.gov/pubmed/18305442.

[62] Gnepp, Douglas R. "Salivary Gland Tumor 'Wishes' to Add to the Next WHO Tumor Classification: Sclerosing Polycystic Adenosis, Mammary Analogue Secretory Carcinoma, Cribriform Adenocarcinoma of the Tongue and Other Sites, and Mucinous Variant of Myoepithelioma." *Head and Neck Pathology* 8, no. 1 (2014): 42–49. https://doi.org/10.1007/s12105-014-0532-x.

[63] Skálová, Alena, Douglas R. Gnepp, Roderick H. W. Simpson, Jean E. Lewis, Dirk Janssen, Radek Sima, Tomas Vanecek, Silvana Di Palma, and Michal Michal. "Clonal Nature of Sclerosing Polycystic Adenosis of Salivary Glands Demonstrated by Using the Polymorphism of the Human Androgen Receptor (HUMARA) Locus as a Marker." *The American Journal of Surgical Pathology* 30, no. 8 (2006): 939–44. https://doi.org/10.1097/00000478-200608000-00002.

[64] Marques, Rita Canas, and Ana Félix. "Invasive Carcinoma Arising from Sclerosing Polycystic Adenosis of the Salivary Gland." *Virchows Archiv* 464, no. 5 (February 2014): 621–25. https://doi.org/10.1007/s00428-014-1551-4.

[65] McHugh J. B., Visscher D. W., and Barnes E. L. "Update on Selected Salivary Gland Neoplasms." *Arch Pathol Lab Med.* 133(11), no.

1763-74 (November 2009). https://doi.org/10.1043/1543-2165-133.11.1763.

[66] Pinkston, John A., and Philip Cole. "Incidence Rates of Salivary Gland Tumors: Results from a Population-Based Study." *Otolaryngology–Head and Neck Surgery* 120, no. 6 (1999): 834–40. https://doi.org/10.1016/s0194-5998(99)70323-2.

[67] Mchugh, Catherine H., Dianna B. Roberts, Adel K. El-Naggar, Ehab Y. Hanna, Adam S. Garden, Merrill S. Kies, Randal S. Weber, and Michael E. Kupferman. "Prognostic Factors in Mucoepidermoid Carcinoma of the Salivary Glands." *Cancer* 118, no. 16 (2011): 3928–36. https://doi.org/10.1002/cncr.26697.

[68] Loh, Kwok Seng, Emma Barker, Guillem Bruch, Brian Osullivan, Dale H. Brown, David P. Goldstein, Ralph W. Gilbert, Patrick J. Gullane, and Jonathan C. Irish. "Prognostic Factors in Malignancy of the Minor Salivary Glands." *Head & Neck* 31, no. 1 (2009): 58–63. https://doi.org/10.1002/hed.20924.

[69] Spiro, Ronald H., Howard T. Thaler, Wesley F. Hicks, Uma A. Kher, Andrew H. Huvos, and Elliot W. Strong. "The Importance of Clinical Staging of Minor Salivary Gland Carcinoma." *The American Journal of Surgery* 162, no. 4 (1991): 330–36. https://doi.org/10.1016/0002-9610(91)90142-z.

[70] Poorten, Vincent L. M. Vander, Alfonsus J. M. Balm, Frans J. M. Hilgers, I. Bing Tan, Ronald B. Keus, and Augustinus A. M. Hart. "Stage as Major Long Term Outcome Predictor in Minor Salivary Gland Carcinoma." *Cancer* 89, no. 6 (2000): 1195–1204. https://doi.org/10.1002/1097-0142(20000915)89:6<1195::aid-cncr2>3.3.co;2-a.

[71] Luna, Mario A. "Salivary Mucoepidermoid Carcinoma: Revisited." *Advances in Anatomic Pathology* 13, no. 6 (2006): 293–307. https://doi.org/10.1097/01.pap.0000213058.74509.d3.

[72] Kaszuba, Scott M., Mark E. Zafereo, David I. Rosenthal, Adel K. El-Naggar, and Randal S. Weber. "Effect of Initial Treatment on Disease Outcome for Patients with Submandibular Gland Carcinoma." *Archives of Otolaryngology–Head & Neck Surgery* 133, no. 6 (January 2007): 546. https://doi.org/10.1001/archotol.133.6.546.

[73] Ellis, Gary L., and Paul L. Auclair. *Atlas of Tumor Pathology.* Washington, D.C., 1996.

[74] Thompson, Lester D., Muhammad N. Aslam, Jennifer N. Stall, Aaron M. Udager, Simion Chiosea, and Jonathan B. Mchugh. "Clinicopathologic and Immunophenotypic Characterization of 25 Cases of Acinic Cell Carcinoma with High-Grade Transformation." *Head and Neck Pathology* 10, no. 2 (June 2015): 152–60. https://doi.org/10.1007/s12105-015-0645-x.

[75] Nagao, Toshitaka. "'Dedifferentiation' and High-Grade Transformation in Salivary Gland Carcinomas." *Head and Neck Pathology* 7, no. S1 (2013): 37–47. https://doi.org/10.1007/s12105-013-0458-8.

[76] Mannelli, Giuditta, Lorenzo Cecconi, Martina Fasolati, Roberto Santoro, Alessandro Franchi, and Oreste Gallo. "Parotid Adenoid Cystic Carcinoma: Retrospective Single Institute Analysis." *American Journal of Otolaryngology* 38, no. 4 (2017): 394–400. https://doi.org/10.1016/j.amjoto.2017.03.008.

[77] Zhang, Chun-Ye, Rong-Hui Xia, Jing Han, Bing-Shun Wang, Wei-Dong Tian, Lai-Ping Zhong, Zhen Tian, Li-Zhen Wang, Yu-Hua Hu, and Jiang Li. "Adenoid Cystic Carcinoma of the Head and Neck: Clinicopathologic Analysis of 218 Cases in a Chinese Population." *Oral Surgery, Oral Medicine, Oral Pathology and Oral Radiology* 115, no. 3 (2013): 368–75. https://doi.org/10.1016/j.oooo.2012.11.018.

[78] Seethala, Raja R., Jennifer L. Hunt, Zubair W. Baloch, Virginia A. Livolsi, and E. Leon Barnes. "Adenoid Cystic Carcinoma with High-Grade Transformation." *The American Journal of Surgical Pathology* 31, no. 11 (2007): 1683–94. https://doi.org/10.1097/pas.0b013e3180dc928c.

[79] Perez, Danyel Elias Da Cruz, Fábio De Abreu Alves, Inês Nobuko Nishimoto, Oslei Paes De Almeida, and Luiz Paulo Kowalski. "Prognostic Factors in Head and Neck Adenoid Cystic Carcinoma." *Oral Oncology* 42, no. 2 (2006): 139–46. https://doi.org/10.1016/j.oraloncology.2005.06.024.

[80] Perzin, Karl H., Patrick Gullane, and Albert C. Clairmont. "Adenoid Cystic Carcinomas Arising in Salivary Glands. A Correlation of Histologic Features and Clinical Course." *Cancer* 42, no. 1 (1978): 265–82. https://doi.org/10.1002/1097-0142(197807)42:1<265::aid-cncr2820420141>3.0.co;2-z.

[81] Szanto, Philip A., Mario A. Luna, M. Eugenia Tortoledo, and Robert A. White. "Histologic Grading of Adenoid Cystic Carcinoma of the Salivary Glands." *Cancer* 54, no. 6 (1984): 1062–69. https://doi.org/10.1002/1097-0142(19840915)54:6<1062::aid-cncr2820540622>3.0.co;2-e.

[82] Chau, Yuk-Ping, Tadashi Hongyo, Katsuyuki Aozasa, and John K. C. Chan. "Dedifferentiation of Adenoid Cystic Carcinoma: Report of a Case Implicating p53 Gene Mutation." *Human Pathology* 32, no. 12 (2001): 1403–7. https://doi.org/10.1053/hupa.2001.28966.

[83] Cheuk, W., John K. C. Chan, and Roger K. C. Ngan. "Dedifferentiation in Adenoid Cystic Carcinoma of Salivary Gland." *The American Journal of Surgical Pathology* 23, no. 4 (1999): 465–72. https://doi.org/10.1097/00000478-199904000-00012.

[84] Ide, F, K Mishima, and I Saito. "Small Foci of High-Grade Carcinoma Cells in Adenoid Cystic Carcinoma Represent an Incipient Phase of Dedifferentiation." *Histopathology* 43, no. 6 (2003): 604–6. https://doi.org/10.1111/j.1365-2559.2003.01682.x.

[85] Moles MA, Avila IR, Archilla AR. Dedifferentiation occurring in adenoid cystic carcinoma of the tongue. *Oral Surg Oral Med Oral Pathol Oral Radiol Endod.* 1999;88:177–180.

[86] Sato, Katsuaki, Yoshimichi Ueda, Aya Sakurai, Yoshimaro Ishikawa, Sachiko Kaji, Takayuki Nojima, and Shogo Katsuda. "Adenoid Cystic Carcinoma of the Maxillary Sinus with Gradual Histologic Transformation to High-Grade Adenocarcinoma: a Comparative Report with Dedifferentiated Carcinoma." *Virchows Archiv* 448, no. 2 (2005): 204–8. https://doi.org/10.1007/s00428-005-0054-8.

[87] Spiro, Ronald H., Andrew G. Huvos, and Elliot W. Strong. "Adenoid Cystic Carcinoma of Salivary Origin." *The American Journal of*

Surgery 128, no. 4 (1974): 512–20. https://doi.org/10.1016/0002-9610(74)90265-7.

[88] Weert, Stijn Van, Isaäc Van Der Waal, Birgit I. Witte, C. René Leemans, and Elisabeth Bloemena. "Histopathological Grading of Adenoid Cystic Carcinoma of the Head and Neck: Analysis of Currently Used Grading Systems and Proposal for a Simplified Grading Scheme." *Oral Oncology* 51, no. 1 (2015): 71–76. https://doi.org/10.1016/j.oraloncology.2014.10.007.

[89] "Adenoid Cystic Carcinoma of the Head and Neck — a 20 Years Experience." *British Dental Journal* 196, no. 10 (2004): 621–21. https://doi.org/10.1038/sj.bdj.4811293.

[90] Garden, Adam S., Randal S. Weber, William H. Morrison, K.kian Ang, and Lester J. Peters. "The Influence of Positive Margins and Nerve Invasion in Adenoid Cystic Carcinoma of the Head and Neck Treated with Surgery and Radiation." *International Journal of Radiation Oncology*Biology*Physics* 32, no. 3 (1995): 619–26. https://doi.org/10.1016/0360-3016(95)00122-f.

[91] Fordice, Jim, Corey Kershaw, Adel El-Naggar, and Helmuth Goepfert. "Adenoid Cystic Carcinoma of the Head and Neck." *Archives of Otolaryngology–Head & Neck Surgery* 125, no. 2 (January 1999): 149. https://doi.org/10.1001/archotol.125.2.149.

[92] Gondivkar, Shailesh M., Amol R. Gadbail, Revant Chole, and Rima V. Parikh. "Adenoid Cystic Carcinoma: A Rare Clinical Entity and Literature Review." *Oral Oncology* 47, no. 4 (2011): 231–36. https://doi.org/10.1016/j.oraloncology.2011.01.009.

[93] Palma, Silvana Di. "Carcinoma Ex Pleomorphic Adenoma, with Particular Emphasis on Early Lesions." *Head and Neck Pathology* 7, no. S1 (2013): 68–76. https://doi.org/10.1007/s12105-013-0454-z.

[94] Eisele, David W., Steven J. Wang, and Lisa A. Orloff. "Electrophysiologic Facial Nerve Monitoring during Parotidectomy." *Head & Neck*, 2009. https://doi.org/10.1002/hed.21190.

[95] Lowry, Thomas R., Thomas J. Gal, and Joseph A. Brennan. "Patterns of Use of Facial Nerve Monitoring During Parotid Gland Surgery."

Otolaryngology–Head and Neck Surgery 133, no. 3 (2005): 313–18. https://doi.org/10.1016/j.otohns.2005.03.010.

[96] Hopkins, C., S. Khemani, R. M. Terry, and D. Golding-Wood. "How We Do It: Nerve Monitoring in ENT Surgery: Current UK Practice." *Clinical Otolaryngology* 30, no. 2 (December 2005): 195–98. https://doi.org/10.1111/j.1365-2273.2004.00933.x.

[97] O'Regan, Barry, Girish Bharadwaj, and Andrew Elders. "Techniques for Dissection of the Facial Nerve in Benign Parotid Surgery: a Cross Specialty Survey of Oral and Maxillofacial and Ear Nose and Throat Surgeons in the UK." *British Journal of Oral and Maxillofacial Surgery* 46, no. 7 (2008): 564–66. https://doi.org/10.1016/j.bjoms.2008.01.008.

[98] Beck, Douglas L., James S. Atkins, James E. Benecke, and Derald E. Brackmann. "Intraoperative Facial Nerve Monitoring: Prognostic Aspects during Acoustic Tumor Removal." *Otolaryngology–Head and Neck Surgery* 104, no. 6 (1991): 780–82. https://doi.org/10.1177/019459989110400602.

[99] Bhattacharyya, Neil, Marc E. Richardson, and Laverne D. Gugino. "An Objective Assessment of the Advantages of Retrograde Parotidectomy." *Otolaryngology–Head and Neck Surgery* 131, no. 4 (2004): 392–96. https://doi.org/10.1016/j.otohns.2004.03.012.

[100] Haenggeli A., Richter M., Lehmann W., and Dulguerov P. "A Complication of Intraoperative Facial Nerve Monitoring: Facial Skin Burns." *Am J Otol.* 20(5):679-82 (September 1999). https://www.ncbi.nlm.nih.gov/pubmed/10503594.

[101] Terrell, J. E., P. R. Kileny, C. Yian, R. M. Esclamado, C. R. Bradford, M. S. Pillsbury, and G. T. Wolf. "Clinical Outcome of Continuous Facial Nerve Monitoring During Primary Parotidectomy." Archives of Otolaryngology - Head and Neck Surgery 123, no. 10 (January 1997): 1081–87. https://doi.org/10.1001/archotol.1997.01900100055008.

[102] López M., Quer M., León X., Orús C., Recher K., and Vergés J. "Usefulness of Facial Nerve Monitoring during Parotidectomy." *Acta*

Otorrinolaringol Esp. 52(5) (July 2001): 418–21. https://www. ncbi.nlm.nih.gov/pubmed/11526649.

[103] Makeieff, Marc, Frdric Venail, C??sar Cartier, Renaud Garrel, Louis Crampette, and Bernard Guerrier. "Continuous Facial Nerve Monitoring during Pleomorphic Adenoma Recurrence Surgery." *The Laryngoscope* 115, no. 7 (2005): 1310–14. https://doi.org/10.1097/ 01.mlg.0000166697.48868.8c.

[104] Quer, M., O. Guntinas-Lichius, F. Marchal, V. Vander Poorten, D. Chevalier, X. León, D. Eisele, and P. Dulguerov. "Classification of Parotidectomies: a Proposal of the European Salivary Gland Society." *European Archives of Oto-Rhino-Laryngology* 273, no. 10 (October 2016): 3307–12. https://doi.org/10.1007/s00405-016-3916-6.

[105] Klintworth, Nils, Johannes Zenk, Michael Koch, and Heinrich Iro. "Postoperative Complications after Extracapsular Dissection of Benign Parotid Lesions with Particular Reference to Facial Nerve Function." *The Laryngoscope* 120, no. 3 (2010): 484–90. https://doi. org/10.1002/lary.20801.

[106] Zernial, Oliver, Ingo N. Springer, Patrick Warnke, Franz Härle, Christian Risick, and Jörg Wiltfang. "Long-Term Recurrence Rate of Pleomorphic Adenoma and Postoperative Facial Nerve Paresis (in Parotid Surgery)." *Journal of Cranio-Maxillofacial Surgery* 35, no. 3 (2007): 189–92. https://doi.org/10.1016/j.jcms.2007.02.003.

[107] Guntinas-Lichius, Orlando, J Peter Klussmann, Claus Wittekindt, and Eberhard Stennert. "Parotidectomy for Benign Parotid Disease at a University Teaching Hospital: Outcome of 963 Operations." *The Laryngoscope* 116, no. 4 (2006): 534–40. https://doi.org/10.1097/01. mlg.0000200741.37460.ea.

[108] Mantsopoulos, Konstantinos, Michael Koch, Nils Klintworth, Johannes Zenk, and Heinrich Iro. "Evolution and Changing Trends in Surgery for Benign Parotid Tumors." *The Laryngoscope* 125, no. 1 (2014): 122–27. https://doi.org/10.1002/lary.24837.

[109] Olsen, Kerry D., Miquel Quer, Remco De Bree, Vincent Vander Poorten, Alessandra Rinaldo, and Alfio Ferlito. "Deep Lobe

Parotidectomy—Why, When, and How?" *European Archives of Oto-Rhino-Laryngology* 274, no. 12 (December 2017): 4073–78. https://doi.org/10.1007/s00405-017-4767-5.

[110] Lim, Y. C., S. Y. Lee, and E. C. Choie. "PD.249 Conservative Parotidectomy for Thetreatment of Parotid Cancers." *Oral Oncology Supplement 1*, no. 1 (2005): 142–43. https://doi.org/10.1016/s1744-7895(05)80366-9.

[111] Huang, Gang, Guangqi Yan, Xinli Wei, and Xin He. "Superficial Parotidectomy versus Partial Superficial Parotidectomy in Treating Benign Parotid Tumors." *Oncology Letters* 9, no. 2 (2014): 887–90. https://doi.org/10.3892/ol.2014.2743.

[112] "*Controversies in the Management of Salivary Gland Disease*," 2012. https://doi.org/10.1093/med/9780199578207.001.0001.

[113] Mantsopoulos, Konstantinos, Michael Koch, Nils Klintworth, Johannes Zenk, and Heinrich Iro. "Evolution and Changing Trends in Surgery for Benign Parotid Tumors." *The Laryngoscope* 125, no. 1 (2014): 122–27. https://doi.org/10.1002/lary.24837.

[114] Mendelsohn, Abie H., Sunita Bhuta, Thomas C. Calcaterra, Hubert B. Shih, Elliot Abemayor, and Maie A. St. John. "Parapharyngeal Space Pleomorphic Adenoma: A 30-Year Review." *The Laryngoscope* 119, no. 11 (2009): 2170–74. https://doi.org/10.1002/lary.20496.

[115] Guntinas-Lichius, Orlando, Jens Peter Klussmann, Ulla Schroeder, Gero Quante, Markus Jungehuelsing, and Eberhard Stennert. "Primary Parotid Malignoma Surgery in Patients With Normal Preoperative Facial Nerve Function: Outcome and Long-Term Postoperative Facial Nerve Function." *The Laryngoscope* 114, no. 5 (2004): 949–56. https://doi.org/10.1097/00005537-200405000-00032.

[116] Otsuka, Kuninori, Yorihisa Imanishi, Yuichiro Tada, Daisuke Kawakita, Satoshi Kano, Kiyoaki Tsukahara, Akira Shimizu, et al. "Clinical Outcomes and Prognostic Factors for Salivary Duct Carcinoma: A Multi-Institutional Analysis of 141 Patients." *Annals of Surgical Oncology* 23, no. 6 (2016): 2038–45. https://doi.org/10.1245/s10434-015-5082-2.

[117] Lombardi, Davide, Marc Mcgurk, Vincent Vander Poorten, Marco Guzzo, Remo Accorona, Vittorio Rampinelli, and Piero Nicolai. "Surgical Treatment of Salivary Malignant Tumors." *Oral Oncology* 65 (2017): 102–13. https://doi.org/10.1016/j.oraloncology. 2016.12.007.

[118] Olsen, Kerry D., Miquel Quer, Remco De Bree, Vincent Vander Poorten, Alessandra Rinaldo, and Alfio Ferlito. "Deep Lobe Parotidectomy—Why, When, and How?" *European Archives of Oto-Rhino-Laryngology* 274, no. 12 (December 2017): 4073–78. https://doi.org/10.1007/s00405-017-4767-5.

[119] Larian, Babak. "Parotidectomy for Benign Parotid Tumors." *Otolaryngologic Clinics of North America* 49, no. 2 (2016): 395–413. https://doi.org/10.1016/j.otc.2015.10.006.

[120] "Book Review Surgery and Disease of the Mouth and Jaws. A Practical Treatise on the Surgery and Diseases of the Mouth and Allied Structures. By Vilroy Papin Blair, AM, MD, Professor of Oral Surgery in the Washington University Dental School, and Associate in Surgery in the Washington University Medical School. St Louis: C. V. Mosby Company. 1912." *The Boston Medical and Surgical Journal* 169, no. 3 (1913): 95–95. https://doi.org/10.1056/nejm1913 07171690319.

[121] Bailey, Hamilton. "The Treatment of Tumours of the Parotid Gland with Special Reference to Total Parotidectomy." *British Journal of Surgery* 28, no. 111 (1941): 337–46. https://doi.org/10.1002/bjs.18 002811102.

[122] Foustanos, Andreas, and Harris Zavrides. "Face-Lift Approach Combined with a Superficial Musculoaponeurotic System Advancement Flap in Parotidectomy." *British Journal of Oral and Maxillofacial Surgery* 45, no. 8 (2007): 652–55. https://doi.org/10. 1016/j.bjoms.2007.05.008.

[123] Casani, A. P., N. Cerchiai, I. Dallan, V. Seccia, And S. Sellari Franceschini. "Benign Tumours Affecting the Deep Lobe of the Parotid Gland: How to Select the Optimal Surgical Approach." *Acta*

Otorhinolaryngol Ital 35(2) (2015): 80–87. https://www.ncbi.nlm. nih.gov/pmc/articles/PMC4443562/.

[124] Gates G. A., Johns M. E. (1980) Embryology and anatomy of the salivary glands. In: Paparella MM, Shumrick DA (eds) Otolaryngology. Saunders, Philadelphia, p 12). https://entokey.com/anatomy-and-physiology-of-the-salivary-glands/.

[125] Saha, Somnath, Sudipta Pal, Moushumi Sengupta, Kanishka Chowdhury, Vedula Padmini Saha, and Lopamudra Mondal. "Identification of Facial Nerveduringg Parotidectomy: A Combined Anatomical & Surgical Study." *Indian Journal of Otolaryngology and Head & Neck Surgery* 66, no. 1 (2013): 63–68. https://doi.org/10.1007/s12070-013-0669-z.

[126] Wong, Wai Keat, and Subhaschandra Shetty. "Classification of Parotidectomy: a Proposed Modification to the European Salivary Gland Society Classification System." *European Archives of Oto-Rhino-Laryngology* 274, no. 8 (November 2017): 3175–81. https://doi.org/10.1007/s00405-017-4581-0.

[127] Klussmann, J. P., T. Ponert, R. P. Mueller, H. P. Dienes, and O. Guntinas-Lichius. "Patterns of Lymph Node Spread and Its Influence on Outcome in Resectable Parotid Cancer." *European Journal of Surgical Oncology* (EJSO) 34, no. 8 (2008): 932–37. https://doi.org/10.1016/j.ejso.2008.02.004.

[128] Pasion, S. L. "The Indications for Elective Treatment of the Neck in Cancer of the Major Salivary Glands." *Journal of Oral and Maxillofacial Surgery* 50, no. 12 (1992): 1347–48. https://doi.org/10.1016/0278-2391(92)90252-u.

[129] Armstrong, John G., Louis B. Harrison, Howard T. Thaler, Hamutal Friedlander-Klar, Daniel E. Fass, Michael J. Zelefsky, Jatin P. Shah, Elliot W. Strong, and Ronald H. Spiro. "The Indications for Elective Treatment of the Neck in Cancer of the Major Salivary Glands." *Cancer* 69, no. 3 (January 1992): 615–19. https://doi.org/10.1002/1097-0142(19920201)69:3<615::aid-cncr2820690303>3.0.co;2-9.

[130] Kawata, Ryo, Lee Koutetsu, Katsuhiro Yoshimura, Shuji Nishikawa, and Hiroshi Takenaka. "Indication for Elective Neck Dissection for

N0 Carcinoma of the Parotid Gland: a Single Institutions 20-Year Experience." *Acta Oto-Laryngologica*, 2009, 1–7. https://doi.org/10.1080/00016480903062160.

[131] Bradley, Patrick J. "Primary Malignant Parotid Epithelial Neoplasm." *Current Opinion in Otolaryngology & Head and Neck Surgery* 23, no. 2 (2015): 91–98. https://doi.org/10.1097/moo.0000000000000139.

[132] Yoo, Shin-Hyuk, Jong-Lyel Roh, Seon-Ok Kim, Kyung-Ja Cho, Seung-Ho Choi, Soon Yuhl Nam, and Sang Yoon Kim. "Patterns and Treatment of Neck Metastases in Patients with Salivary Gland Cancers." *Journal of Surgical Oncology* 111, no. 8 (2015): 1000–1006. https://doi.org/10.1002/jso.23914.

[133] Terhaard, Chris H. j., Herman Lubsen, Coen R. N. Rasch, Peter C. Levendag, Hans H.à.m. Kaanders, Reineke E. Tjho-Heslinga, Piet L. A. Van Den Ende, and Fred Burlage. "The Role of Radiotherapy in the Treatment of Malignant Salivary Gland Tumors." *International Journal of Radiation Oncology*Biology*Physics* 61, no. 1 (2005): 103–11. https://doi.org/10.1016/j.ijrobp.2004.03.018.

[134] Laramore, G. E., John M Krall, Thomas W Griffin, William Duncan, Melvin P Richter, Kurubarahalli R Saroja, Moshe H Maor, and Lawrence W Davis. "Neutron versus Photon Irradiation for Unresectable Salivary Gland Tumors: Final Report of an RTOG-MRC Randomized Clinical Trial." *International Journal of Radiation Oncology*Biology*Physics* 27, no. 2 (1993): 235–40. https://doi.org/10.1016/0360-3016(93)90233-1.

[135] Hsieh, Cheng-En, Chien-Yu Lin, Li-Yu Lee, Lan-Yan Yang, Chun-Chieh Wang, Hung-Ming Wang, Joseph Tung-Chieh Chang, et al. "Adding Concurrent Chemotherapy to Postoperative Radiotherapy Improves Locoregional Control but Not Overall Survival in Patients with Salivary Gland Adenoid Cystic Carcinoma—a Propensity Score Matched Study." *Radiation Oncology* 11, no. 1 (2016). https://doi.org/10.1186/s13014-016-0617-7.

[136] Keller, Gunter, Diana Steinmann, Alexander Quaas, Viktor Grünwald, Stefan Janssen, and Kais Hussein. "New Concepts of Personalized Therapy in Salivary Gland Carcinomas." *Oral Oncology*

68 (2017): 103–13. https://doi.org/10.1016/j.oraloncology.2017.02.018.
[137] Wang, Xiaoli, Yijun Luo, Minghuan Li, Hongjiang Yan, Mingping Sun, and Tingyong Fan. "Management of Salivary Gland Carcinomas - a Review." *Oncotarget* 8, no. 3 (2016). https://doi.org/10.18632/oncotarget.13952.
[138] Marchese-Ragona, R., C. De Filippis, G. Marioni, and A. Staffieri. "Treatment of Complications of Parotid Gland Surgery." *Acta Otorhinolaryngol Ital.* 25(3) (June 2005): 174–78. https://www.ncbi.nlm.nih.gov/pmc/articles/PMC2639867/.
[139] Bailey BJ. *Head and Neck Surgery-Otolaryngology. 3rd Edn.* Philadelphia, PA: Lippincott Williams & Wilkins; 2001.
[140] Dulguerov, Pavel, Francis Marchal, and Willy Lehmann. "Postparotidectomy Facial Nerve Paralysis: Possible Etiologic Factors and Results with Routine Facial Nerve Monitoring." *The Laryngoscope* 109, no. 5 (1999): 754–62. https://doi.org/10.1097/00005537-199905000-00014.
[141] Reilly, J. "Facial Nerve Stimulation and Postparotidectomy Facial Paresis." *Otolaryngology - Head and Neck Surgery* 128, no. 4 (2003): 530–33. https://doi.org/10.1016/s0194-5998(03)00089-5.
[142] Hui, Yau, David S. Y. Wong, Ling-Yuen Wong, Wai-Kuen Ho, and William I Wei. "A Prospective Controlled Double-Blind Trial of Great Auricular Nerve Preservation at Parotidectomy." *The American Journal of Surgery* 185, no. 6 (2003): 574–79. https://doi.org/10.1016/s0002-9610(03)00068-0.
[143] Kerawala, Cyrus, Peter A. Brennan, Luke Cascarini, Darryl Godden, Darryl Coombes, and Jim Mccaul. "Management of Tumour Spillage during Parotid Surgery for Pleomorphic Adenoma." *British Journal of Oral and Maxillofacial Surgery* 52, no. 1 (2014): 3–6. https://doi.org/10.1016/j.bjoms.2013.05.143.
[144] Park, Gi Cheol, Kyung-Ja Cho, Jun Kang, Jong-Lyel Roh, Seung-Ho Choi, Sang Yoon Kim, and Soon Yuhl Nam. "Relationship between Histopathology of Pleomorphic Adenoma in the Parotid Gland and Recurrence after Superficial Parotidectomy." *Journal of Surgical*

Oncology 106, no. 8 (2012): 942–46. https://doi.org/10.1002/jso. 23202.

[145] Laskawi, Rainer, Thomas Schott, Maritta Mirzaie-Petri, and Michael Schroeder. "Surgical Management of Pleomorphic Adenomas of the Parotid Gland: A Followup Study of Three Methods." *Journal of Oral and Maxillofacial Surgery* 54, no. 10 (1996): 1176–79. https://doi.org/10.1016/s0278-2391(96)90344-4.

[146] Witt, Robert L., David W. Eisele, Randall P. Morton, Piero Nicolai, Vincent Vander Poorten, and Peter Zbären. "Etiology and Management of Recurrent Parotid Pleomorphic Adenoma." *The Laryngoscope* 125, no. 4 (July 2014): 888–93. https://doi.org/10. 1002/lary.24964.

[147] 147. Orabona, Giovanni Dellaversana, Paola Bonavolontà, Giorgio Iaconetta, Raimondo Forte, and Luigi Califano. "Surgical Management of Benign Tumors of the Parotid Gland: Extracapsular Dissection versus Superficial Parotidectomy—Our Experience in 232 Cases." *Journal of Oral and Maxillofacial Surgery* 71, no. 2 (2013): 410–13. https://doi.org/10.1016/j.joms.2012.05.003.

[148] Barzan, Luigi, and Marco Pin. "Extra-Capsular Dissection in Benign Parotid Tumors." *Oral Oncology* 48, no. 10 (2012): 977–79. https://doi.org/10.1016/j.oraloncology.2012.05.010.

[149] Olsen, Kerry D. "Superficial Parotidectomy." *Operative Techniques in General Surgery* 6, no. 2 (2004): 102–14. https://doi.org/10.1053/ j.optechgensurg.2004.05.006.

[150] 152-138. Piekarski, Janusz, Dariusz Nejc, Wiesław Szymczak, Konrad Wroński, and Arkadiusz Jeziorski. "Results of Extracapsular Dissection of Pleomorphic Adenoma of Parotid Gland." *Journal of Oral and Maxillofacial Surgery* 62, no. 10 (2004): 1198–1202. https://doi.org/10.1016/j.joms.2004.01.025.

[151] Laccourreye, Henri, Ollivier Laccourreye, R??gis Cauchois, V??ronique Jouffre, Madeleine M??nard, and Daniel Brasnu. "Total Conservative Parotidectomy for Primary Benign Pleomorphic Adenoma of the Parotid Gland." *The Laryngoscope* 104, no. 12 (1994). https://doi.org/10.1288/00005537-199412000-00011.

[152] Hui, Yau, David S. Y. Wong, Ling-Yuen Wong, Wai-Kuen Ho, and William I Wei. "A Prospective Controlled Double-Blind Trial of Great Auricular Nerve Preservation at Parotidectomy." *The American Journal of Surgery* 185, no. 6 (2003): 574–79. https://doi.org/10.1016/s0002-9610(03)00068-0.

[153] Forfar, A. r. "Sternocleidomastid Muscle Transfer and Superificial Musculoaponeurotic System Plication in the Prevention of Freys Syndrome." *Journal of Oral and Maxillofacial Surgery* 49, no. 8 (1991): 912. https://doi.org/10.1016/0278-2391(91)90032-h.

[154] Rhee, John S., Richard E. Davis, and W. Jarrard Goodwin. "Minimizing Deformity from Parotid Gland Surgery." *Current Opinion in Otolaryngology & Head and Neck Surgery* 7, no. 2 (1999): 90. https://doi.org/10.1097/00020840-199904000-00010.

[155] Bjerkhoel, A., and O. Trobbe. "Freys Syndrome: Treatment with Botulinum Toxin." *The Journal of Laryngology & Otology* 111, no. 9 (1997): 839–44. https://doi.org/10.1017/s0022215100138769.

[156] Laccourreye, Ollivier, Elie Akl, Raimundo Gutierrez-Fonseca, Dominique Garcia, Daniel Brasnu, and Brigitte Bonan. "Recurrent Gustatory Sweating (Frey Syndrome) After Intracutaneous Injection of Botulinum Toxin Type A." *Archives of Otolaryngology–Head & Neck Surgery* 125, no. 3 (January 1999): 283. https://doi.org/10.1001/archotol.125.3.283.

[157] Laskawi, Rainer, Christian Drobik, and Claudia Schönebeck. "Up-to-Date Report of Botulinum Toxin Type A Treatment in Patients With Gustatory Sweating (Freys Syndrome)." *The Laryngoscope* 108, no. 3 (1998): 381–84. https://doi.org/10.1097/00005537-199803000-00013.

[158] Restivo, D. A., S. Lanza, F. Patti, S. Giuffrida, R. Marchese-Ragona, P. Bramanti, and A. Palmeri. "Improvement of Diabetic Autonomic Gustatory Sweating by Botulinum Toxin Type A." *Neurology* 59, no. 12 (2002): 1971–73. https://doi.org/10.1212/01.wnl.0000036911.75478.fa.

[159] Tugnoli, V., R. Marchese Ragona, R. Eleopra, R. Quatrale, J. G. Capone, A. Pastore, C. Montecucco, and D. De Grandis. "The Role of

Gustatory Flushing in Freys Syndrome and Its Treatment with Botulinum Toxin Type A." *Clinical Autonomic Research* 12, no. 3 (January 2002): 174–78. https://doi.org/10.1007/s10286-002-0026-x.

[160] Zinis, Luca Oscar Redaelli De, Michela Piccioni, Antonino Roberto Antonelli, and Piero Nicolai. "Management and Prognostic Factors of Recurrent Pleomorphic Adenoma of the Parotid Gland: Personal Experience and Review of the Literature." *European Archives of Oto-Rhino-Laryngology* 265, no. 4 (2007): 447–52. https://doi.org/10.1007/s00405-007-0502-y.

[161] Dulguerov, Pavel, Jelena Todic, Marc Pusztaszeri, and Naif H. Alotaibi. "Why Do Parotid Pleomorphic Adenomas Recur? A Systematic Review of Pathological and Surgical Variables." *Frontiers in Surgery* 4 (2017). https://doi.org/10.3389/fsurg.2017.00026.

[162] Witt, Robert L., David W. Eisele, Randall P. Morton, Piero Nicolai, Vincent Vander Poorten, and Peter Zbären. "Etiology and Management of Recurrent Parotid Pleomorphic Adenoma." *The Laryngoscope* 125, no. 4 (July 2014): 888–93. https://doi.org/10.1002/lary.24964.

[163] Wittekindt, Claus, Kristina Streubel, Georg Arnold, Eberhard Stennert, and Orlando Guntinas-Lichius. "Recurrent Pleomorphic Adenoma of the Parotid Gland: Analysis of 108 Consecutive Patients." *Head & Neck* 29, no. 9 (2007): 822–28. https://doi.org/10.1002/hed.20613.

[164] Henriksson, Gert, Karl Magnus Westrin, Bengt Carls��, and Claes Silfversw�Rd. "Recurrent Primary Pleomorphic Adenomas of Salivary Gland Origin." *Cancer* 82, no. 4 (1998): 617–20. https://doi.org/10.1002/(sici)1097-0142(19980215)82:4<617::aid-cncr1>3.0.co;2-i.

[165] Riad, Magdy Amin, Hussein Abdel-Rahman, Waleed F. Ezzat, Ahmad Adly, Ossama Dessouky, and Mohamed Shehata. "Variables Related to Recurrence of Pleomorphic Adenomas: Outcome of Parotid Surgery in 182 Cases." *The Laryngoscope* 121, no. 7 (October 2011): 1467–72. https://doi.org/10.1002/lary.21830.

[166] Krolls, Sigurds O., and Robert C. Boyers. "Mixed Tumors of Salivary Glands." *Cancer* 30, no. 1 (1972): 276–81. https://doi.org/10.1002 /1097-0142(197207)30:1<276::aid-cncr2820300138>3.0.co;2-v.

[167] Zbären, Peter, Isabelle Tschumi, Michel Nuyens, and Edouard Stauffer. "Recurrent Pleomorphic Adenoma of the Parotid Gland." *The American Journal of Surgery* 189, no. 2 (2005): 203–7. https://doi.org/10.1016/j.amjsurg.2004.11.008.

[168] Glas, Afina S., Albert Vermey, Harry Hollema, Peter H. Robinson, Jan L. N. Roodenburg, Raoul E. Nap, and John Th. M. Plukker. "Surgical Treatment of Recurrent Pleomorphic Adenoma of the Parotid Gland: A Clinical Analysis of 52 Patients." *Head & Neck* 23, no. 4 (2001): 311–16. https://doi.org/10.1002/hed.1036.

[169] Makeieff, Marc, Pierfrancesco Pelliccia, Flavie Letois, Grégoire Mercier, Sebastien Arnaud, Cartier César, Renaud Garrel, Louis Crampette, and Bernard Guerrier. "Recurrent Pleomorphic Adenoma: Results of Surgical Treatment." *Annals of Surgical Oncology* 17, no. 12 (2010): 3308–13. https://doi.org/10.1245/s10434-010-1173-2.

[170] Stennert, Eberhard, Claus Wittekindt, Jens Peter Klussmann, Georg Arnold, and Orlando Guntinas-Lichius. "Recurrent Pleomorphic Adenoma of the Parotid Gland: A Prospective Histopathological and Immunohistochemical Study." *The Laryngoscope* 114, no. 1 (2004): 158–63. https://doi.org/10.1097/00005537-200401000-00030.

[171] Abu-Ghanem, Yasmin, Aviram Mizrachi, Aron Popovtzer, Nora Abu-Ghanem, and Raphael Feinmesser. "Recurrent Pleomorphic Adenoma of the Parotid Gland: Institutional Experience and Review of the Literature." *Journal of Surgical Oncology* 114, no. 6 (2016): 714–18. https://doi.org/10.1002/jso.24392.

[172] Liu, Huawei, Weisheng Wen, Haitao Huang, Yongqiang Liang, Xinying Tan, Sanxia Liu, Changkui Liu, and Min Hu. "Recurrent Pleomorphic Adenoma of the Parotid Gland." *Otolaryngology–Head and Neck Surgery* 151, no. 1 (2014): 87–91. https://doi.org/10.1177/ 0194599814528098.

[173] Samson, Michael J., Ralph Metson, C. C. Wang, and William W. Montgomery. "Preservation of the Facial Nerve in the Management

of Recurrent Pleomorphic Adenoma." *The Laryngoscope* 101, no. 10 (1991). https://doi.org/10.1288/00005537-199110000-00006.

[174] Douglas, James G., John Einck, Mary Austin-Seymour, Wui-Jin Koh, and George E. Laramore. "Neutron Radiotherapy for Recurrent Pleomorphic Adenomas of Major Salivary Glands." *Head & Neck* 23, no. 12 (2001): 1037–42. https://doi.org/10.1002/hed.10027.

[175] Renehan, Andrew, E. Neville Gleave, and Mark Mcgurk. "An Analysis of the Treatment of 114 Patients with Recurrent Pleomorphic Adenomas of the Parotid Gland." *The American Journal of Surgery* 172, no. 6 (1996): 710–14. https://doi.org/10.1016/s0002-9610(96)00293-0.

[176] Carew, John F., Ronald H. Spiro, Bhuvanesh Singh, and Jatin P. Shah. "Treatment of Recurrent Pleomorphic Adenomas of the Parotid Gland." *Otolaryngology–Head and Neck Surgery* 121, no. 5 (1999): 539–42. https://doi.org/10.1016/s0194-5998(99)70053-7.

[177] Myssiorek, David, Carlos B. Ruah, and Roger L. Hybels. "Recurrent Pleomorphic Adenomas of the Parotid Gland." *Head & Neck* 12, no. 4 (1990): 332–36. https://doi.org/10.1002/hed.2880120410.

[178] Chen, Allen M., Joaquin Garcia, M. Kara Bucci, Jeanne M. Quivey, and David W. Eisele. "Recurrent Pleomorphic Adenoma of the Parotid Gland: Long-Term Outcome of Patients Treated with Radiation Therapy." *International Journal of Radiation Oncology*Biology*Physics* 66, no. 4 (2006): 1031–35. https://doi.org/10.1016/j.ijrobp.2006.06.036.

[179] Wallace, Audrey S., Christopher G. Morris, Jessica M. Kirwan, John W. Werning, and William M. Mendenhall. "Radiotherapy for Pleomorphic Adenoma." *American Journal of Otolaryngology* 34, no. 1 (2013): 36–40. https://doi.org/10.1016/j.amjoto.2012.08.002.

[180] Samson, Michael J., Ralph Metson, C. C. Wang, and William W. Montgomery. "Preservation of the Facial Nerve in the Management of Recurrent Pleomorphic Adenoma." *The Laryngoscope* 101, no. 10 (1991). https://doi.org/10.1288/00005537-199110000-00006.

[181] Niparko, J. K., M. L. Beauchamp, C. J. Krause, S. R. Baker, and W. P. Work. "Surgical Treatment of Recurrent Pleomorphic Adenoma of

the Parotid Gland." *Archives of Otolaryngology - Head and Neck Surgery* 112, no. 11 (January 1986): 1180–84. https://doi.org/10.1001/archotol.1986.03780110056007.

[182] Dawson, A. K. "Radiation Therapy in Recurrent Pleomorphic Adenoma of the Parotid." *International Journal of Radiation Oncology*Biology*Physics* 16, no. 3 (1989): 819–21. https://doi.org/10.1016/0360-3016(89)90501-4.

BIOGRAPHICAL SKETCH

Salma Mohammed Al Sheibani, MD

Affiliation: Department of ENT, Al Nahdha Hospital, Muscat, Sultanate of Oman.

Education:

Scholarship in Otolaryngology University of Pittsburgh and School of Medicine, United states, 2008
Surgical Oncology Training (Tata Memorial Center, Mumbia, India), 2007
Arab Board of Otolaryngology and Head & Neck surgery, 2004
Oman Medical Board Specialization, 2006
MD, College of Medicine, SQU, Sultanate of Oman, 1999
Bachelor of Science (Health Science), SQU, Sultanate of Oman, 1996.

Business Address: ENT department, Al Nahdha Hospital, Muscat, Oman.

Research and Professional Experience:

- Involved actively in departmental and national research.

- Working in a tertiary care center, managing all head and neck cancer and complex airway problems from whole country.
- President of ENT society since April 2019.
- Focal point of head and neck services of ENT departments in Oman.
- Assistant program director of Oman Medical Specialty Board since January 2018.
- Chairman of examination subcommittee in OMSB since 2012.
- Member of examination board at Oman Medical Specialty Board since June 2019.
- Chairman of quality improvement committee at Al Nahdha Hospital , 2011-2012.
- Focal point of ENT at the public education services at Ministry of Health.
- Member in American head and neck society since 2013.
- A reviewer in Oman Medical Journal and SQU Journal.
- Examiner at Arab Board of medical specialization since 2015.
- A member of examination subcommittee at Arab Board of medical specialization since January 2018.

Professional Appointments:

- Senior consultant in otolaryngology and head and neck surgery since 2008.
- Head of ENT department at Al Nahdha Hospital since 2017.
- Assistant program director of Oman Medical Specialty Board since January 2018.
- A senior clinical lecturer at Sultan Qaboos University, College of Medicine & Health Sciences since January 2011.
- A trainer in Oman Medical Specialty Board since 2009

Honors:

- Best staff award, Al Nahdha Hospital, 2013 and 2019

- Best research award (Management of parotid neoplasm), At Al Nahdha Hospital, 2016
- Best presentation Award (Outcome of head and neck cancer Omani national screening campaign), GCC conference Dubai 2017
- Best trainer award, OMSB, 2012.

Publications from the Last 3 Years:

1. Al-Sheibani, Salma M., Kiran P. Sawardekar, Salwa J. Habib, and Hunaina M. Al-Kindi. "Nasopharyngeal Salivary Gland Anlage Tumour: A Rare Cause of Neonatal Respiratory Distress." *Sultan Qaboos University Medical Journal (SQUMJ)* 18, no. 2 (September 2018): 211. https://doi.org/10.18295/squmj.2018.18.02.015.
2. Salma M., Fahad B., Faisal K., and Jamil H. "Managment of Parotid Neoplasm, Al Nahdha Experience." *Submitted to Oman Medical Journal.*
3. Salma M., and Faisal K. "Pediatric sialoendoscopy Our Experience." *Submitted to Oman Medical Journal.*
4. Yahya S., Salma M., Amer W., and Faisal K. "Influence of Recurrent Laryngeal Nerve Injury on the Occurrence of Laryngopharyngeal Reflux, a Preliminary Study." *Submitted to Oman Medical Journal.*
5. Zaina D., Salma M., Faisal K., and Jamil H. "Outcome of Head and Neck Cancer National Screening Campaign." *Submitted to Sultan Qaboos University Medical Journal.*

In: Salivary Glands　　　　　　　　　ISBN: 978-1-53617-497-7
Editor: Alicia S. Bryant　　　　　　　© 2020 Nova Science Publishers, Inc.

Chapter 3

JUVENILE RECURRENT PAROTITIS

Salma M. Al Sheibani*, MD
ENT department, Al Nahdha Hospital, Muscat, Sultanate of Oman

ABSTRACT

Juvenile recurrent parotitis is the second most common salivary gland disease in children after mumps. It's more common in males and in the age group between 3 and 6 years. It is characterized by recurrent attacks of non-suppurative and non-obstructive parotid inflammation. The pathogenesis is multifactorial and therefore the management is a challenge due to etiological diversity.

The main criterion for establishing the severity is the frequency of attacks. The treatment of the acute phase is with antibiotics and analgesics. Juvenile recurrent parotitis usually resolves spontaneously after puberty, however, in some cases the disease may continue leading to progressive loss of parenchymal functions. Subsequently, it may lead to a major intervention such as parotidectomy.

The diagnosis is established by history and clinical examination. Various imaging modalities are also used such as ultrasound, sialography and MR sialography.

---

* Corresponding Author Email: salsheibani@gmail.com.

The main difficulty of treatment of this disease concerns the prevention of the recurrence of inflammatory episodes.

Various medical and surgical measures have been used, but none of them proved to be useful in preventing or treating the attacks of juvenile recurrent parotitis. Sialoendoscopy is used as a diagnostic and therapeutic modality in juvenile recurrent parotitis. The main aim of this treatment is to reduce the recurrent attacks of parotitis and prevent irreversible changes in the parotid glands by irrigating and dilating the ductal system of the parotid gland under direct vision. Another advantage is the opportunity to inject medications intraductally under direct vision. Various intraductal lavage solutions were effective in prevention of recurrence regardless of their composition as it breaks the inflammatory cycle.

This chapter reviews the clinical presentation, etiology and diagnostic modality of juvenile recurrent parotitis. In addition to the review of acute juvenile recurrent parotitis attacks management, the role of sialoendoscopy in the diagnosis, management and prevention of recurrent attacks is addressed. Moreover, various solutions that can be used as a lavage solution and their role in the management of juvenile recurrent parotitis will be reviewed.

Keywords: juvenile recurrent parotitis, sialendoscopy, recurrent acute parotitis

INTRODUCTION

Juvenile recurrent parotitis is defined as a recurrent non-suppurative and non-obstructive parotid inflammation. It is the second most common salivary gland disease in children after mumps. It is more common in males and in the age group between 3 and 6 years [1].

PATHOPHYSIOLOGY

The pathogenesis is multifactorial and various etiologies have been proposed such as altered salivary composition, intraductal malformations, allergies, viral infections, immune disorders, particularly IgA deficiency and a familial form with autosomal inheritance [2, 3]. These etiologies lead to salivary changes that result in ascending oral bacteria through Stenson's duct leading to chronic infection and dilation of the distal ducts [3]. Furthermore, dehydration and an impaired salivary flow might cause low grade inflammation of the gland and ductal epithelium, with subsequent stricture and columnar metaplasia. This will lead to an increase in mucous secretion, resulting in decreased clearance of the more viscous saliva and further reduction in salivary flow, predisposing patients to recurrent parotitis [3]. These sequels are seen on imaging and histological examination as a dilatation of the distal ducts of the parotid gland and punctuate sialectasis with subsequent chronic inflammation of the glandular parenchyma.

Chronic inflammation of the gland leads to a lymphocytic periductal and intralobular infiltrate, which can have cytotoxic effects on the glandular parenchyma, leading to parenchymal destruction with the development of sialectasis predisposing to recurrence [4].

CLINICAL PRESENTATION

Each attack of parotitis is usually associated with intermittent painful swelling of one or both glands, often accompanied by local erythema and fever (Figure 1) [1].

Symptoms are most often one-sided; in case of bilateral involvement, the disease appears to be significantly more symptomatic on one side.

The main criterion for establishing the severity is the frequency of attacks. Each episode usually lasts for a few days and may occur every 2-3 months [5].

Juvenile recurrent parotitis usually resolves spontaneously after puberty, however, in some cases the disease may continue leading to progressive loss of the parenchymal function. Recurrences influence the quality of life and can also lead to progressive glandular destruction and may consequently lead to major intervention such as parotidectomy [5]. The prevention of recurrent attacks represents the most important aspect of this pathology.

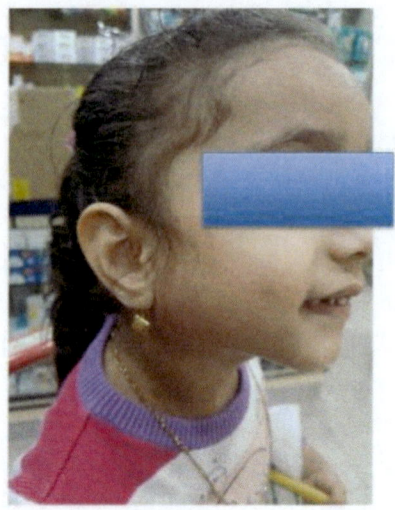

Figure 1. A 4-year-old girl presented with right recurrent parotitis. She underwent sialoendoscope that showed typical findings of juvenile recurrent parotitis.

DIAGNOSIS

Juvenile recurrent parotitis must be differentiated from sialolithiasis and other causes of unilateral parotid swelling such as tumors, vascular malformations, branchial cleft cyst or lymphadenopathy. Ultrasound, sialography, CT scan, MRI and MR sialography can be used. CT scan and sialography should be avoided due to radiation exposure. Sialography is no longer performed in most of the countries due to the evolution of other imaging techniques [6, 7]. MRI is performed in selected cases, as it requires sedation and takes a long time to perform. Acute inflammation appears as a hypointense area in T1-weighted MRI and hyperintense in T2-weighted image. On the other hand, chronic inflammation is seen as an isointense area on both T1- and T2-weighted sequences. MR Sialography provides good visualization of the ductal system, taking advantage of particular T2-weighted sequences [8-11].

Ultrasound is the imaging of choice in the pediatric population as it does not involve radiation and does not require sedation. The typical sonographic findings in juvenile recurrent parotitis are heterogenous glands with multiple small, rounded hypoechoic and anechoic areas corresponding to sialectasia and enlarged intraglandular lymph nodes [6, 12]. Sometimes diffuse microcalcifications are seen in the parotid gland which should be differentiated from inflammatory parotid disease. If it is sialolithiasis, the microclacifications are seen in the duct and are large compared to the small multiple parenchymal microcalcifications seen in inflammatory disease [16, 13].

TREATMENT

The management is a challenge due to the diversity in etiologies.

The treatment of an acute phase constitutes a combination of sialogogues, parotid gland massage and broad-spectrum antibiotics [14].

Various medical and surgical measures have been used in order to induce glandular atrophy. However, none of them have proved to be useful in preventing the attacks of juvenile recurrent parotitis or treating it.

Sialoendoscopy is a novel procedure that has modified the diagnosis and treatment of salivary gland diseases, especially juvenile recurrent parotitis [9]. It is a minimally invasive procedure that allows endoscopic visualization of the salivary ductal system. The two main indications of doing it in children are: 1) a minimum of two episodes of parotitis occurring within 6 months [15-17] or 2) two episodes occurring within one year [18-20].

SIALOENDOSCOPY

Sialoendoscopy is used as a diagnostic and therapeutic modality in juvenile recurrent parotitis.

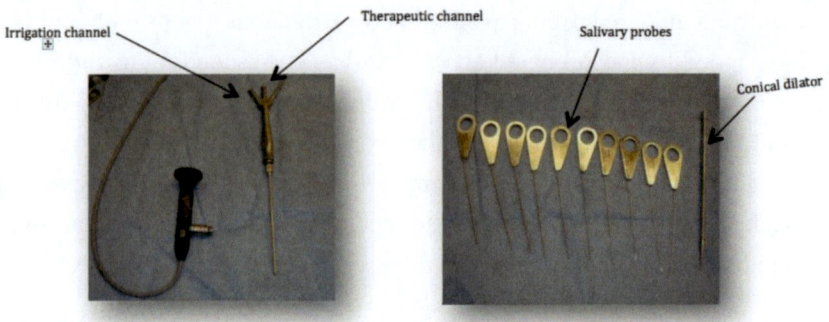

Figure 2. Sialoendscopy set.

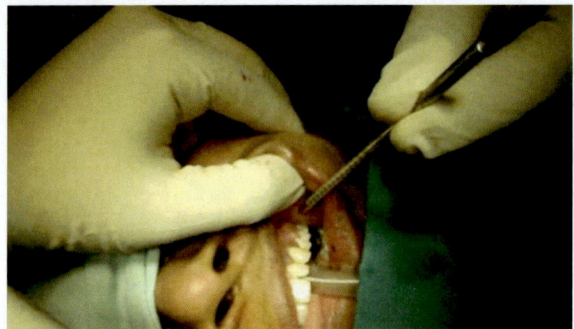

Figure 3. Papilla dilatation using conical dilator.

Technique:

Marchal all in one 1.1 or 1.3 mm sialendoscopes are used for diagnostic or interventional sialendoscopy (Figure 2) [21].

In children the procedure is performed under general anesthesia. The papilla is carefully dilated using a conical dilator (Figure 3). Then, the duct is dilated using duct probes. The probes are increased gradually in diameters starting at size 0000 until reaching size 3 (Figure 4). After the insertion of a size 3 probe, the duct lumen is large enough to accommodate the sialoendoscope.

Sialoendoscopy is performed with continuous rinsing of normal saline that flows around the optic fiber in order to improve the visualization of the ductal system and to flush out all the debris and mucus plaques that can interfere with visualization of the ductal system.

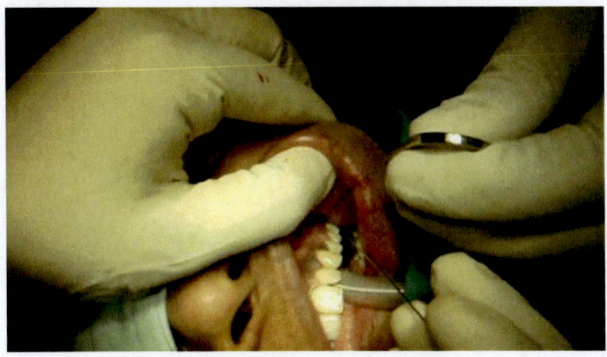

Figure 4. Duct dilatation using salivary probes.

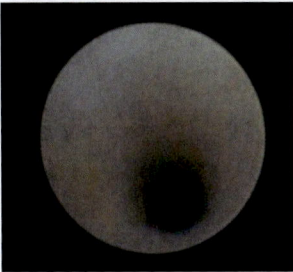

Figure 5. The typical sialoendoscopic findings in juvenile recurrent parotitis of whitish, avascular and stenotic duct.

If there is a stenosis, then serial endoscopic and bougies (with guide wire guidance) dilation is performed followed by dexamethasone (8 mg) intraductal injection.

Post-operatively, the patient is advised to do gland massage and take amoxicillin with clavulanic acid antibiotic and nonsteroidal anti-inflammatory drugs for 7 days.

The most commonly recognized diagnostic sialendoscopic finding is whitish, avascular and stenotic Stenson's duct Figure 5 [1]. The absence of a natural vascularization may affect the sphincteral system of the parotid gland and reduces the ability to drain saliva [22]. Sialendoscopy breaks the cycle of inflammation by washing out intraductal debris and dilating stenosis due to pressure of irrigating fluid that result in the reduction of the recurrent attacks of parotitis and preventing the irreversible changes in the parotid glands [4, 23, 24].

Another advantage of doing sialoendoscopy is the opportunity to inject intraductal medications under direct vision. The therapeutic efficacy of sialoendoscopy is mostly due to tissue irrigation and clearing out the debris regardless of the solution used [4, 23, 24].

SILAOENDOSCOPIC TREATMENT OUTCOME

Sialoendoscopic intervention is promising and it may result in complete symptom resolution or reduction in the frequency of attacks. Nahlieli et al. (2004) [22], in a series of 26 cases of JRP treated by dilatation and abundant washing, a resolution of symptoms was observed in 92% of the cases. A similar study done by Quenin et al. (2009) [4] who assessed 10 children with symptomatic JRP and initial ultrasound evaluation revealed a white duct without vascularity. The sialendoscopy success rate is reported as 89%. In 2009, Shacham et al. [18] included 70 children with JRP who were treated with sialendoscopic dilatation and lavage (saline followed by hydrocortisone) with a success rate of 93%.

The meta-analysis conducted by Canzi et al. (2013) [24], based on 10 trials comprising 179 patients, showed a very low recurrence rate after sialendoscopy and lavage (14%). Moreover, the meta-analysis by Ramakrishna et al. (2015) [25], based on 7 studies with 120 patients and 165 glands, reported similar results. The primary success rate of interventional sialendoscopy was 73% (95% CI [64-82]). These two meta-analyses demonstrated that sialoendoscopy had low morbidity and no major complications encountered.

The earlier the sialoendoscopic intervention was performed, the more likely that the symptoms would resolve after a single sialoendoscopic intervention. This breaks the inflammatory cycle and reduces the cytotoxic glandular damage. The delay in the treatment of juvenile recurrent parotitis resulted in irreversible ductal cystic changes and the duct appears like a sausage shape (Figure 6 A and B). In such cases, a complete resolution of symptoms by sialoendoscopic intervention can never be achieved.

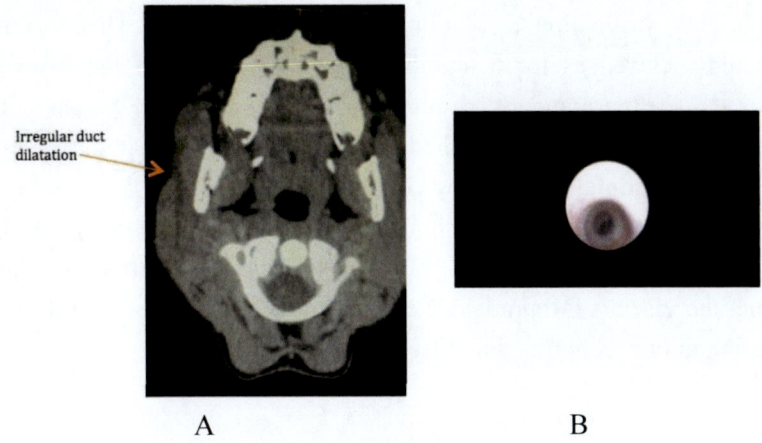

Figure 6. (A and B) A. A CT scan of a 10 year old girl with juvenile recurrent parotitis demonstrating irregular duct dilatation; B. Sialoendoscopic view demonstrating a sausage shape (cystic dilatation and constriction) duct which is irreversible.

SIALOENDOSCOPY LAVAGE SOLUTIONS

Intraductal lavage breaks the vicious cycle of decreased secretion, stasis and infection by evacuating mucus plugs and intraductal debris. Various lavage solutions have been reported with successful outcomes such as, hydrocortisone, physiological saline, antibiotic, Lipiodol or a combination of any of these solutions [18, 26-28].

Galili et al. (1986) [29] were the first authors to report the benefits of intraductal lavage in the treatment of JRP. Their study, based on a series of 22 children, demonstrated a marked reduction of acute recurrences after intraductal lavage during diagnostic lipiodol sialography. This beneficial effect was attributed to the procedure. Very good results were also described by Katz et al. (2009) [26] in the treatment of 840 children diagnosed between 2000 and 2007. These children were treated at the acute phase by oral antibiotic therapy consisting of a macrolide (spiramycin) and a nitroimidazole (metronidazole) in combination with an antispasmodic (phloroglu-cinol) and corticosteroid therapy at the dose of 1 mg/kg for two weeks. Lipiodol sialography was performed in the case of recurrent parotitis

(753 patients), following resolution of the acute episode. This treatment eliminated recurrences for 6 months to 1 year in 95% of the cases and symptoms were permanently resolved in two thirds of the patients. This study, although retrospective, is the largest study to be published to date and supported the findings reported by Galili et al. (1986) [29].

Intraductal lavage performed with or without sialoendoscopy was very effective as shown by Roby et al. (2015) [30]. The most important thing is breaking the viscous inflammatory cycle of juvenile recurrent parotitis and preventing the irreversible glandular damage.

SIALOENDOSCOPY COMPLICATIONS

Sialoendoscopy is generally a safe procedure, however, complications may be encountered in 2-25% of the cases [16, 17, 18, 20, 31-37].

The pediatric sialendoscopy complications are as follows:

1. Transient gland swelling: this is an expected event and considered part of the procedure rather than a complication. It occurs due to excessive saline irrigation during the procedure. It will stay from a few hours to few days. It is usually resolved with gland massage [32].
2. Duct perforation: there are two signs to identify this complication: a) hazy sialoendoscopic appearance (Figure 7) and/or b) progressive gland swelling. This complication can be managed by stent insertion bypassing the site of the perforated duct [36].
3. Stenosis: this may develop as a result of trauma to the duct or orifice during the procedure. It is a preventable complication and managed by inserting a stent and intraductal injection of hydrocortisone. In case of complete stenosis gland removal, excision of the stenotic segment and vein graft or botulinum toxin injection should be taken into consideration [32].
4. Duct avulsion: this mandates gland removal or botulinum toxin injection [36].

5. Airway obstruction: this is due to swelling of the deep part of the gland as a result of excessive irrigation. It is an avoidable complication prevented by controlling the amount of fluid used for irrigation and if it happened, it can be treated by intravenous injection of hydrocortisone and gland massage [31, 32, 36].

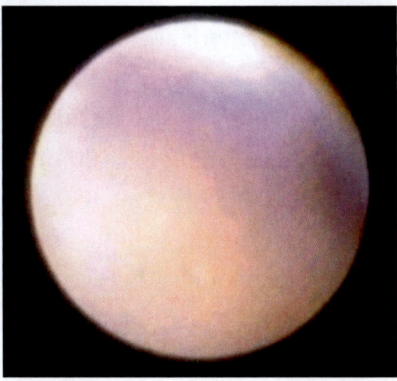

Figure 7. Sialoendoscopic image demonstrating a hazy image that is suggestive of a perforation.

CONCLUSION

Juvenile recurrent parotitis is a recurrent non-suppurative and non-obstructive parotid inflammation. It is more commonly seen in boys. Multiple etiologies have been proposed with multifactorial pathology that makes management challenging.

The treatment of the acute phase is based on a combination of sialogogues, parotid gland massage and broad-spectrum antibiotics.

Sialendoscopy is a novel, minimally invasive procedure that can be used as a diagnostic and therapeutic option. Its therapeutic efficacy is related to the inflammation suppression by washing out the intraductal debris and dilating the stenosis. The earlier the sialoendoscopic intervention is performed, the less likely is the glandular damage and more likely that the patient will benefit from single intervention.

REFERENCES

[1] Reid, E., F. Douglas, Y. Crow, A. Hollman, and J. Gibson. "Autosomal Dominant Juvenile Recurrent Parotitis." *Journal of Medical Genetics* 35, no. 5 (January 1998): 417-19. https://doi.org/ 10.1136/jmg. 35.5.417.

[2] Ericson, Sune, Birgitta Zetterlund, and Jan Öhman. "Recurrent Parotitis and Sialectasis in Childhood: Clinical, Radiologic, Immunologic, Bacteriologic, and Histologic Study." *Annals of Otology, Rhinology & Laryngology* 100, no. 7 (1991): 527-35. https://doi.org/10.1177/000348949110000702.

[3] Fazekas, Tamas, Peter Wiesbauer, Brigitte Schroth, Ulrike, Potschger, Helmut Gadner, and Andreas Heitger. "Selective IgA Deficiency in Children with Recurrent Parotitis of Childhood." *The Pediatric Infectious Disease Journal* 24, no. 5 (2005): 461-62. https://doi.org/10.1097/01.inf.0000160994.65328.dd.

[4] Quenin, Stephanie, Isabelle Plouin-Gaudon, Francis Marchal, Patrick Froehlich, Francois Disant, and Frederic Faure. "Juvenile Recurrent Parotitis." *Archives of Otolaryngology - Head & Neck Surgery* 134, no. 7 (2008): 715. https://doi.org/10.1001/archotol.134.7.715.

[5] Leerdam, Cm, Hco Martin, and D. Isaacs. "Recurrent Parotitis of Childhood." *Journal of Paediatrics and Child Health* 41, no. 12 (2005): 631-34. https://doi.org/10.1111/j.1440-1754.2005.00773.x.

[6] Ogden, Margaret A., Kristina W. Rosbe, and Jolie L. Chang. "Pediatric Sialendoscopy Indications and Outcomes." *Current Opinion in Otolaryngology & Head and Neck Surgery* 24, no. 6 (2016): 529-35. https://doi.org/10.1097/moo.0000000000000314.

[7] Chung, Man Ki, Han-Sin Jeong, Moon-Hee Ko, Hyun-Jin Cho, Nam-Gyu Ryu, Do-Yeon Cho, Young-Ik Son, and Chung-Hwan Baek. "Pediatric Sialolithiasis: What Is Different from Adult Sialolithiasis?" *International Journal of Pediatric Otorhinolaryngology* 71, no. 5 (2007): 787-91. https://doi.org/10. 1016/j.ijporl.2007.01.019.

[8] Gadodia, Ankur, Ashu Seith, Raju Sharma, and Alok Thakar. "MRI and MR Sialography of Juvenile Recurrent Parotitis." *Pediatric Radiology* 40, no. 8 (2010): 1405-10. https://doi.org/10.1007/s00247-010-1639-1.

[9] Huisman, Thierry A. G. M., David Holzmann, and David Nadal. "MRI of Chronic Recurrent Parotitis in Childhood." *Journal of Computer Assisted Tomography* 25, no. 2 (2001): 269-73. https://doi.org/10.1097/00004728-200103000-00021.

[10] Berlucchi, Marco, Vittorio Rampinelli, Marco Ferrari, Paola Grazioli, and Luca O. Redaelli De Zinis. "Sialoendoscopy for Treatment of Juvenile Recurrent Parotitis: The Brescia Experience." *International Journal of Pediatric Otorhinolaryngology* 105 (2018): 163-66. https://doi.org/10.1016/j.ijporl.2017.12.024.

[11] Mandel, Louis, and Renuka Bijoor. "Imaging (Computed Tomography, Magnetic Resonance Imaging, Ultrasound, Sialography) in a Case of Recurrent Parotitis in Children." *Journal of Oral and Maxillofacial Surgery* 64, no. 6 (2006): 984-88. https://doi.org/10.1016/j.joms.2005.11.057.

[12] Shimizu, Mayumi, Jürgen Ußmüller, Karl Donath, Kazunori Yoshiura, Shigeo Ban, Shigenobu Kanda, Satoru Ozeki, and Masanori Shinohara. "Sonographic Analysis of Recurrent Parotitis in Children." *Oral Surgery, Oral Medicine, Oral Pathology, Oral Radiology, and Endodontology* 86, no. 5 (1998): 606-15. https://doi.org/10.1016/s1079-2104(98)90355-9.

[13] Rosbe, Kristina W., Dimiter Milev, and Jolie L. Chang. "Effectiveness and Costs of Sialendoscopy in Pediatric Patients with Salivary Gland Disorders." *The Laryngoscope* 125, no. 12 (2015): 2805-9. https://doi.org/10.1002/lary.25384.

[14] Mikolajczak, Stefanie, Moritz Friedo Meyer, Dirk Beutner, and Jan Christoffer Luers. "Treatment of Chronic Recurrent Juvenile Parotitis Using Sialendoscopy." *Acta Oto-Laryngologica* 134, no. 5 (June 2014): 531-35. https://doi.org/10.3109/00016489.2013.879738.

[15] Papadopoulou-Alataki, E., Chatziavramidis, A., Vampertzi, O., Alataki, S., Konstantinidis, I., and "Evaluation and Management of

Juvenile Recurrent Parotitis in Children from Northern Greece." *Hippokratia* 19(4) (October 2015): 356-59. https://www.ncbi.nlm.nih.gov/pubmed/27688702.

[16] Walvekar, Rohanr, Evelyna Kluka, Celeste Gary, and Barry Schaitkin. "Interventional Sialendoscopy for Treatment of Juvenile Recurrent Parotitis." *Journal of Indian Association of Pediatric Surgeons* 16, no. 4 (2011): 132. https://doi.org/10.4103/0971-9261.86865.

[17] Konstantinidis, I., A. Chatziavramidis, E. Tsakiropoulou, H. Malliari, and J. Constantinidis. "Pediatric Sialendoscopy under Local Anesthesia: Limitations and Potentials." *International Journal of Pediatric Otorhinolaryngology* 75, no. 2 (2011): 245-49. https://doi.org/10.1016/j.ijporl.2010.11.009.

[18] Shacham, Rachel, Eitan Bar Droma, Daniel London, Tal Bar, and Oded Nahlieli. "Long-Term Experience with Endoscopic Diagnosis and Treatment of Juvenile Recurrent Parotitis." *Journal of Oral and Maxillofacial Surgery* 67, no. 1 (2009): 162-67. https://doi.org/10.1016/j.joms.2008.09.027.

[19] Capaccio, P., P. E. Sigismund, N. Luca, P. Marchisio, and L. Pignataro. "Modern Management of Juvenile Recurrent Parotitis." *The Journal of Laryngology & Otology* 126, no. 12 (2012): 1254-60. https://doi.org/10.1017/s0022215112002319.

[20] Ardekian, L., H. Klein, R. Al Abri, and F. Marchal. "Sialendoscopy for the Diagnosis and Treatment of Juvenile Recurrent Parotitis." *Revue De Stomatologie, De Chirurgie Maxillo-Faciale Et De Chirurgie Orale* 115, no. 1 (2014): 17-21. https://doi.org/10.1016/j.revsto.2013.12.005.

[21] Faure F., Boem A., Taffin C., Badot F., Disant F, "Marchal F. Sialendoscopie diagnostique et interventionnelle." *Rev. Stomatol. Chir. Maxillofac.* 106, no 4 (2005): 250-252. https://www.sciencedirect.com/science/article/pii/S0035176805858541.

[22] Nahlieli, O. "Juvenile Recurrent Parotitis: A New Method of Diagnosis and Treatment." *Pediatrics* 114, no. 1 (January 2004): 9-12. https://doi.org/10.1542/peds.114.1.9.

[23] Jabbour, Noel, Robert Tibesar, Timothy Lander, and James Sidman. "Sialendoscopy in Children." *International Journal of Pediatric Otorhinolaryngology* 74, no. 4 (2010): 347-50. https://doi.org/10.1016/j.ijporl.2009.12.013.

[24] Canzi P., Occhini A., Pagella F., "Sialendoscopy in Juvenile Recurrent Parotitis: A Review of yhe Literature." *Acta Otorhinolaryngol. Ital.* 33, no 6 (2013): 367-373. https://www.ncbi.nlm.nih.gov/pmc/articles/PMC3870450/.

[25] Ramakrishna, Jayant, Julie Strychowsky, Michael Gupta, and Doron D. Sommer. "Sialendoscopy for the Management of Juvenile Recurrent Parotitis: A Systematic Review and Meta-Analysis." *The Laryngoscope* 125, no. 6 (2014): 1472-79. https://doi.org/10.1002/lary.25029.

[26] Katz, Philippe, Dana M. Hartl, and Agnès Guerre. "Treatment of Juvenile Recurrent Parotitis." *Otolaryngologic Clinics of North America* 42, no. 6 (2009): 1087-91. https://doi.org/10.1016/j.otc.2009.09.002.

[27] Nahlieli, Oded, Tal Bar, Rachel Shacham, Eli Eliav, and Liat Hecht-Nakar. "Management of Chronic Recurrent Parotitis: Current Therapy." *Journal of Oral and Maxillofacial Surgery* 62, no. 9 (2004): 1150-55. https://doi.org/10.1016/j.joms.2004.05.116.

[28] Faure, Frederic, Patrick Froehlich, and Francis Marchal. "Paediatric Sialendoscopy." *Current Opinion in Otolaryngology & Head and Neck Surgery* 16, no. 1 (2008): 60-63. https://doi.org/10.1097/moo.0b013e3282f45fe1.

[29] Galili, Dan, and Yitzhak Marmary. "Juvenile Recurrent Parotitis: Clinicoradiologic Follow-up Study and the Beneficial Effect of Sialography." *Oral Surgery, Oral Medicine, Oral Pathology* 61, no. 6 (1986): 550-56. https://doi.org/10.1016/0030-4220(86)90091-5.

[30] Roby, Brianne Barnett, Jameson Mattingly, Emily L. Jensen, Dexiang Gao, and Kenny H. Chan. "Treatment of Juvenile Recurrent Parotitis of Childhood." *JAMA Otolaryngology - Head & Neck Surgery* 141, no. 2 (January 2015): 126. https://doi.org/10.1001/jamaoto.2014.3036.

[31] Baptista, P., C. V. Gimeno, F. Salvinelli, V. Rinaldi, and M. Casale. "Acute Upper Airway Obstruction Caused by Massive Oedema of the Tongue: Unusual Complication of Sialoendoscopy." *The Journal of Laryngology & Otology* 123, no. 12 (2009): 1402-3. https://doi.org/10.1017/s0022215109005647.

[32] Nahlieli, Oded. "Advanced Sialoendoscopy Techniques, Rare Findings, and Complications." *Otolaryngologic Clinics of North America* 42, no. 6 (2009): 1053-72. https://doi.org/10.1016/j.otc.2009.08.007.

[33] Schwarz, Yehuda, Aren Bezdjian, and Sam J. Daniel. "Sialendoscopy in Treating Pediatric Salivary Gland Disorders: a Systematic Review." *European Archives of Oto-Rhino-Laryngology* 275, no. 2 (April 2017): 347-56. https://doi.org/10.1007/s00405-017-4830-2.

[34] Semensohn, Ryan, Zorik Spektor, David J. Kay, Alfredo S. Archilla, and David L. Mandell. "Pediatric Sialendoscopy: Initial Experience in a Pediatric Otolaryngology Group Practice." *The Laryngoscope* 125, no. 2 (May 2014): 480-84. https://doi.org/10.1002/lary.24868.

[35] Hackett, Alyssa M., Christopher F. Baranano, Michael Reed, Umamaheswar Duvvuri, Richard J. Smith, and Deepak Mehta. "Sialoendoscopy for the Treatment of Pediatric Salivary Gland Disorders." *Archives of Otolaryngology - Head & Neck Surgery* 138, no. 10 (January 2012): 912. https://doi.org/10.1001/2013.jamaoto.244.

[36] Martins-Carvalho, Christine, Isabelle Plouin-Gaudon, Stéphanie Quenin, Jérome Lesniak, Patrick Froehlich, Francis Marchal, and Frederic Faure. "Pediatric Sialendoscopy." *Archives of Otolaryngology - Head & Neck Surgery* 136, no. 1 (2010): 33. https://doi.org/10.1001/archoto.2009.184.

[37] Faure, Frederic, Stephanie Querin, Pavel Dulguerov, Patrick Froehlich, Francois Disant, and Francis Marchal. "Pediatric Salivary Gland Obstructive Swelling: Sialendoscopic Approach." *The Laryngoscope* 117, no. 8 (2007): 1364-67. https://doi.org/10.1097/mlg.0b013e318068657c.

INDEX

A

acetylcholine, 66
adenocarcinoma, 9, 16, 26, 36, 38, 50, 62
adenoma, 10, 26, 36, 38, 41, 42, 44, 45, 51, 62, 66, 67, 68
adhesions, 52
age, ix, 22, 23, 44, 46, 49, 50, 97, 98
aggressive behavior, 38, 45, 49
anastomosis, 63
angiogenesis, 9, 24, 32
annotation, 15
antibiotic, 103, 105
anticholinergic, 65
anti-inflammatory drugs, 103
antisense, 3
apoptosis, 10, 13, 20, 27
aspiration, viii, 35, 65, 71
assessment, vii, viii, ix, 35, 36, 39, 52
atrophy, 18, 101
autoantigens, 20, 28
autoimmune diseases, 18
autoimmunity, 18

B

bacteria, 99
base, 3, 22, 39, 43
beneficial effect, 105
benefits, 105
benign, viii, 9, 10, 11, 23, 27, 30, 35, 36, 39, 42, 44, 45, 55, 63
benign tumors, 10, 23, 24, 36
biological processes, 6
biomarkers, vii, viii, 2, 3, 4, 23, 30, 31
biopsy, 71
bleeding, 65
body fluid, viii, 2, 3
branching, 5, 6, 7, 28, 29, 31
breast carcinoma, 17, 26

C

cadaver, 59
cancer, viii, 2, 3, 7, 10, 11, 16, 22, 24, 25, 31, 36, 42, 43, 55, 62
cancer cells, 24
cancer progression, 11

capsule, 44, 54, 63, 68
carcinogenesis, 14
carcinoma, 9, 10, 17, 25, 26, 27, 28, 29, 30, 31, 32, 33, 36, 38, 44, 46, 47, 48, 49, 50, 51, 55, 62, 68, 80
cell death, 8, 31
cell invasion, 16
cell line, viii, 2, 4, 13, 16, 17, 19
chemical, 66
chemotherapy, vii, ix, 36, 46, 62, 87
children, ix, 37, 46, 50, 97, 98, 101, 102, 104, 105
circulating biomarkers, vii, viii, 2, 4
classification, viii, 2, 3, 37, 38, 43, 45, 59, 60, 61
clinical examination, x, 97
clinical presentation, vii, x, 38, 98
clinical trials, 63
coding, vii, 1, 2, 3, 27, 30, 32
complications, vii, ix, 35, 36, 53, 63, 64, 104, 106
composition, x, 98
compression, 44, 52
connective tissue, 44
controlled studies, 53
correlation, 8, 9, 12, 22, 26, 53
correlations, 7, 20, 32
corticosteroid therapy, 105
costimulatory molecules, 21
CT scan, viii, 35, 39, 40, 41, 52, 100, 105
cure, 51

D

deficiency, 98
degradation, 8
degradation process, 8
dehydration, 99
depression, 65
deregulation, 10, 21
destruction, 18, 99
detection, 30
developmental process, 5, 6
differential diagnosis, 27
dilation, 99, 103
diseases, viii, 2, 101
diversity, ix, 37, 97, 101
down-regulation, 12
drug delivery, 24
drug resistance, 24
dry eyes, 18

E

ECM degradation, 6
encoding, 10
entrapment, 52
epigenetic modification, 18
epigenetics, 18, 30
epithelial cells, 18, 20, 21, 45
epithelium, 5, 6, 99
Epstein Barr, 46
etiology, vii, x, 18, 46, 98
eukaryotic, vii, 1
evidence, 2, 10, 43, 62, 64
evolution, 100
excision, 40, 45, 50, 54, 55, 57, 106
exposure, 44, 56, 57, 68, 100

F

facelift, 56
facial nerve, viii, ix, 35, 36, 40, 41, 42, 43, 46, 48, 52, 53, 54, 55, 57, 58, 59, 60, 61, 63, 64, 66, 67, 68
facial palsy, 38, 62
fascia, 56
female rat, 45
fetal development, 5
fibrosis, 52, 67, 68
first molar, 6

FNAC, 36, 39, 41, 42
formation, 6, 13, 50
functional analysis, 21
fusion, 16, 17, 25, 27, 47

G

gene expression, viii, 1, 3, 6, 16, 18
general anesthesia, 102
genes, 2, 3, 5, 6, 7, 10, 12, 17, 19, 34
genetic predisposition, 18
genome-wide profiling, viii, 2, 3
gland, viii, ix, x, 2, 4, 5, 6, 8, 15, 17, 18, 19, 21, 22, 25, 26, 27, 28, 29, 30, 31, 35, 36, 37, 42, 46, 47, 51, 63, 98, 99, 103, 106, 107
glycosylation, 28
grades, 38, 47, 62
grading, 38, 46, 49
growth, 5, 6, 11, 14, 17, 25, 27, 30, 33, 38, 48, 49
growth factor, 5, 6

H

haemostasis, 65
head and neck cancer, 94, 95
hemangioma, 38
histological examination, 99
histology, 37, 38, 53
human, viii, 2, 3, 4, 12, 13, 14, 17, 21, 23, 27, 29, 31, 33
hydrocortisone, 104, 105, 106, 107

I

identification, 8, 22, 52, 54, 59, 68
image, 40, 57, 100, 107
imaging modalities, x, 97
immunity, 22

immunohistochemistry, 49
in situ hybridization, 11
in vitro, 6, 11, 13, 14, 16, 17, 28
in vivo, 11, 13, 17
incidence, 36, 44, 45, 48, 49, 50, 51, 53, 54, 62, 64, 67, 68
individuals, 19
infection, 99, 105
inflammation, ix, 18, 97, 98, 99, 100, 103, 107
inflammatory disease, 101
injury, iv, 52, 65, 67
innate immunity, 21, 31
insertion, 59, 102, 106
institutions, 53
interface, 39
intervention, x, 97, 99, 104, 107
invasive lesions, 38
ipsilateral, ix, 36, 43
irrigation, 104, 106, 107
isolation, 23

J

juvenile recurrent parotitis, vii, ix, x, 97, 98, 99, 100, 101, 103, 104, 105, 106, 107

K

Keloid, 65

L

landscape, 13
latency, 45
lead, x, 64, 68, 97, 98, 99
lesions, 13, 20, 38, 63, 64
liver, 16, 49
lncRNAs, vii, 1, 2, 3, 4, 17, 18, 20
localization, 52

long non-coding RNAs (lncRNAs), vii, 1, 2, 3, 4, 17, 18, 20
luciferase, 21
lumen, 102
lymph, 16, 39, 43, 45, 48, 49, 54, 62, 100, 101
lymph node, 16, 39, 43, 45, 48, 49, 54, 62, 101
lymphoma, 38

M

machinery, 10
majority, vii, viii, 1, 2, 10, 22, 35, 38, 39, 47, 51, 52, 64
malignancy, 9, 39, 42, 44, 47, 48, 50, 52, 54
malignant tumors, viii, 2, 23, 39, 54
management, vii, viii, ix, x, 8, 35, 36, 38, 41, 42, 97, 98, 101, 107
mandible, 39, 43, 55
mass, 40, 42, 48, 50, 51, 52, 57
mastoid, 41, 56, 58, 59
measurements, 7
medical, x, 95, 98, 101
mesenchyme, 6, 28
messenger RNA, 6
metastasis, 10, 12, 13, 14, 16, 24, 29, 32, 33, 43, 45, 48, 49, 50, 51, 54, 55, 62
metastatic disease, 47, 62
microRNA, 12, 25, 26, 27, 28, 30, 31, 32, 33, 34
microRNAs (miRNAs), v, vii, 1, 2, 3, 4, 5, 6, 7, 8, 9, 10, 11, 12, 13, 15, 16, 17, 18, 19, 20, 21, 22, 23, 24, 25, 26, 27, 28, 29, 32, 33, 34
migration, 12, 13, 14, 21, 24, 33
molecular biology, 2
morbidity, viii, 35, 67, 104
morphogenesis, 5, 6, 28, 29, 31
mortality rate, 46
MRI, viii, 35, 39, 41, 66, 100, 108, 109

mRNAs, viii, 2, 3, 7, 9, 12, 13, 15, 20, 33
mucin, 28
mucoid, 44
mucosa, 38
mucus, 102, 105
mumps, ix, 97, 98
mutation, 26

N

nanometer, 4, 24
necrosis, 65
neoplasm, 11, 95
nerve, viii, 35, 40, 43, 48, 52, 53, 54, 55, 57, 59, 60, 63, 64, 65, 67, 68
neuroma, 41, 65
nitric oxide, 21
nodes, ix, 16, 36, 43
nodules, 66, 67, 68
normal development, 6
nucleic acid, 4, 24
nucleotides, 3

O

obstruction, 107
old age, 44
oncogenesis, 16
oral antibiotic, 105
organs, 5, 6, 15
overlap, 4

P

pain, 38, 48
palliative, ix, 36, 62, 63
paralysis, 48, 53, 64, 66, 67
parenchyma, 44, 99
paresis, 53, 54

parotid, vii, viii, ix, x, 7, 23, 26, 30, 31, 35, 36, 38, 39, 40, 41, 42, 43, 44, 45, 46, 47, 48, 50, 51, 52, 53, 54, 55, 57, 59, 60, 61, 62, 63, 65, 68, 71, 75, 95, 97, 98, 99, 100, 101, 103, 107
parotid cancer, 36, 43, 62
parotid gland, viii, x, 7, 23, 26, 30, 35, 36, 39, 41, 45, 46, 47, 48, 50, 51, 54, 55, 57, 60, 61, 65, 75, 98, 99, 101, 103, 107
parotid neoplasms, 36, 41
parotid tumor, vii, ix, 36, 38, 40, 41, 42, 44, 52, 54, 55, 57, 63
parotidectomy, x, 36, 42, 46, 53, 54, 55, 60, 66, 67, 69, 73, 75, 81, 82, 83, 84, 85, 88, 89, 97, 99
parotitis, vii, ix, x, 52, 97, 98, 99, 100, 101, 103, 104, 105, 106, 107
pathogenesis, ix, 17, 18, 19, 20, 21, 22, 65, 97, 98
pathology, 32, 66, 99, 107
pathway, 10, 12, 17, 26, 28, 29, 30
perforation, 106, 107
peripheral blood mononuclear cell, 19
phosphorylation, 20
physiology, vii, viii, 2, 4, 85
post-transcriptional regulation, 22
preservation, 46
prevention, vii, x, 98, 99
primary tumor, 43
progenitor cell, 5, 7
prognosis, viii, 2, 12, 23, 25, 47, 48, 49, 51
proliferation, 5, 6, 11, 13, 14, 17, 24, 30, 31
protection, 64
proteins, vii, 1, 2, 4, 15, 24
public education, 95

Q

quality improvement, 95
quality of life, 99
quantification, 6

R

radiation, vii, ix, 36, 44, 46, 52, 63, 66, 68, 100, 101
radiation therapy, ix, 36, 66
radiotherapy, 36, 46, 47, 48, 49, 50, 62, 86, 87, 92, 93
recurrence, ix, x, 12, 36, 44, 45, 46, 48, 49, 51, 64, 66, 68, 98, 99, 104
recurrent acute parotitis, 98
regeneration, 5, 7, 65
regression, 7
regression analysis, 7
relevance, 29
resection, 41, 46, 48, 50, 52, 56, 59, 60, 63, 68
resolution, 104, 106
response, 20, 29, 53
restoration, 13
risk, viii, ix, 11, 35, 36, 42, 48, 54, 64
RNAs, v, vii, 1, 2, 3, 17, 19, 27, 32, 33

S

saliva, viii, 2, 4, 6, 9, 23, 24, 27, 31, 99, 103
salivary gland, vii, viii, ix, 2, 4, 5, 6, 7, 8, 9, 10, 11, 13, 15, 16, 17, 18, 20, 21, 22, 23, 25, 26, 27, 28, 29, 30, 32, 33, 34, 35, 36, 37, 38, 39, 44, 46, 47, 48, 50, 51, 55, 63, 85, 97, 98, 101
salivary gland development and physiology, vii, viii, 2, 4
salivary gland tumor(s), viii, ix, 2, 4, 8, 27, 32, 36, 37, 38, 44, 48, 55, 63
science, 110
senescence, 31
sensitivity, 14, 23, 30, 41
sequencing, 12, 13, 19, 32
sialendoscopy, 98, 102, 103, 104, 106, 107, 108, 109, 110, 111, 112

sialography, x, 97, 100, 105
skin, 43, 48, 55, 56, 65, 67
society, 94, 95
solution, x, 63, 64, 98, 104
specialization, 95
squamous cell carcinoma, 16, 32, 36
stenosis, 103, 106, 107
sternocleidomastoid, 56, 66
stimulation, 53
stretching, 52, 64
suppression, 14, 107
surface area, 6
surgical resection, ix, 36, 42, 44, 45, 48, 53, 56, 64, 66
survival, ix, 8, 12, 14, 17, 27, 36, 47, 48, 49, 50, 55, 62
swelling, 99, 100, 106, 107
syndrome, viii, 2, 4, 18, 28, 29, 30, 31, 32, 33, 34, 54, 65

T

target, 6, 7, 9, 13, 14, 15, 16, 18, 19, 20, 21, 25, 28
targeted therapy, 36
techniques, viii, ix, 2, 36, 56, 100
therapeutic targets, vii, viii, 2, 4, 8, 24, 25
therapy, ix, 15, 25, 36, 62, 105
tissue, viii, 1, 3, 4, 5, 6, 9, 10, 11, 12, 13, 14, 15, 16, 17, 20, 24, 27, 38, 39, 52, 54, 60, 65, 104
toxin, 64, 65, 66, 106
transcription, 3, 7, 17
transcription factors, 4, 7
transformation, 44, 45, 48, 49, 68
trauma, 52, 64, 67, 68, 106

treatment, vii, ix, x, 15, 25, 36, 47, 48, 49, 53, 54, 62, 65, 66, 97, 98, 101, 104, 105, 107
tumor, viii, 2, 3, 6, 9, 10, 11, 13, 14, 15, 16, 17, 24, 29, 36, 38, 39, 40, 41, 42, 43, 44, 45, 47, 48, 49, 50, 51, 52, 53, 54, 55, 56, 57, 62, 63, 64, 67, 68, 75
tumor growth, 10, 11, 13, 15, 17
tumor progression, 14, 24, 29
tumorigenesis, vii, viii, 2, 4, 14, 27
tumors, vii, viii, ix, 2, 4, 8, 11, 14, 17, 18, 23, 27, 30, 32, 35, 36, 37, 38, 44, 45, 46, 47, 50, 51, 54, 55, 57, 62, 63, 100

U

ultrasound, viii, x, 23, 35, 39, 97, 104
underlying mechanisms, 10

V

validation, 11, 12
vascularization, 103
very low density lipoprotein, 7
visualization, 100, 101, 102

W

Washington, 73, 78, 85
World Health Organization, 37, 70, 76

X

xenografts, 13, 14
xerostomia, 18

Related Nova Publications

OTOTOXICITY: SIGNS, SYMPTOMS AND TREATMENT

EDITOR: Gregg Colon

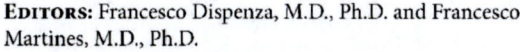

SERIES: Otolaryngology Research Advances

BOOK DESCRIPTION: This compilation focuses on ototoxicity, the adverse pharmacological reaction affecting the inner ear or auditory nerve, characterized by cochlear or vestibular dysfunction. Although ototoxic medications play an essential role in modern medicine, they can cause harm and lead to significant morbidity.

SOFTCOVER ISBN: 978-1-53616-396-4
RETAIL PRICE: $95

SENSORINEURAL HEARING LOSS: PATHOPHYSIOLOGY, DIAGNOSIS AND TREATMENT

EDITORS: Francesco Dispenza, M.D., Ph.D. and Francesco Martines, M.D., Ph.D.

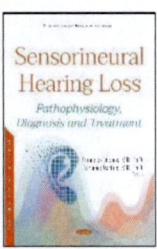

SERIES: Otolaryngology Research Advances

BOOK DESCRIPTION: In this book, the authors present the hearing loss in all its facets, starting from the basis of pathophysiology and anatomy, passing through the clinical and instrumental diagnosis and, finally, describing the most important diseases causing hearing loss with reasonable treatment options.

HARDCOVER ISBN: 978-1-53615-048-3
RETAIL PRICE: $230

To see a complete list of Nova publications, please visit our website at www.novapublishers.com

Related Nova Publications

CRUSH OTOLARYNGOLOGY BOARDS. VOLUME 1

EDITOR: Mohamad R. Chaaban, M.D.

SERIES: Otolaryngology Research Advances

BOOK DESCRIPTION: This book is intended as a study guide for students, residents and practicing otolaryngologists. Volume One includes five sections: General Otolaryngology; Sleep Medicine; Pediatric Otolaryngology; Laryngology; and Rhinology.

HARDCOVER ISBN: 978-1-53614-781-0
RETAIL PRICE: $230

CRUSH OTOLARYNGOLOGY BOARDS. VOLUME 2

EDITOR: Mohamad R. Chaaban, M.D.

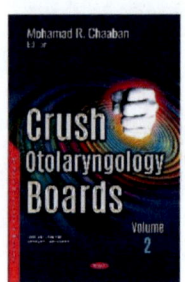

SERIES: Otolaryngology Research Advances

BOOK DESCRIPTION: This book is intended as a study guide for students, residents and practicing otolaryngologists. Volume Two includes the following sections: Otology; Facial Plastics and Reconstructive Surgery; and Head and Neck Cancer.

HARDCOVER ISBN: 978-1-53614-801-5
RETAIL PRICE: $230

To see a complete list of Nova publications, please visit our website at www.novapublishers.com